Frank Norris has a degree in aerospace engineering from the University of Florida. He worked most of his career on the Space Shuttle program, including developing theoretical models to predict Space Shuttle system performance. His easy-to-read style takes us on a journey to explain how colonizing Mars might really occur and why we should care.

Frank Norris

Colonizing Mars

How it Will Happen in our Lifetime

Austin Macauley Publishers™
London * Cambridge * New York * Sharjah

The story, experiences, and words are the author's alone.

Ordering Information
Quantity sales: Special discounts are available on quantity purchases by corporations, associations, and others. For details, contact the publisher at the address below.

Publisher's Cataloging-in-Publication data
Norris, Frank
Colonizing Mars

ISBN 9798889103837 (Paperback)
ISBN 9798889103844 (Hardback)
ISBN 9798889103851 (ePub e-book)

Library of Congress Control Number: 2023913699

www.austinmacauley.com/us

First Published 2023
Austin Macauley Publishers LLC
40 Wall Street, 33rd Floor, Suite 3302
New York, NY 10005
USA

mail-usa@austinmacauley.com
+1 (646) 5125767

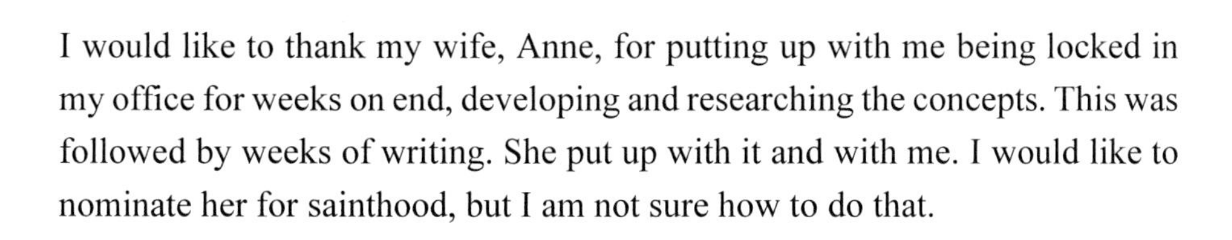
I would like to thank my wife, Anne, for putting up with me being locked in my office for weeks on end, developing and researching the concepts. This was followed by weeks of writing. She put up with it and with me. I would like to nominate her for sainthood, but I am not sure how to do that.

Introduction

I am an aerospace engineer and perhaps a bit of a scientist, though definitely more on the practical side of finding solutions to real problems rather than expanding knowledge where there is no 'problem' that we are facing. I was a math major before switching to engineering. When I was much younger, I dropped out of college for a couple of years but became fascinated with the idea of space travel and the survival of the human species, mostly through the writings of a few scientists and a bit of science fiction. Those who inspired me included Carl Sagan, Isaac Asimov, Robert Heinlein, and others. This led me to go back to college to do what I could to help the space program and perhaps have a small part in saving the human species from the fate of viruses in a limited environment like a Petri dish. (They all die.)

After getting out of college (aerospace engineering degree, University of Florida), I got a job on the Space Shuttle program and worked in various roles for almost 31 years. For the first couple of years, I developed theoretical models to predict performance of space shuttle systems. I often read graduate-level texts to learn new topics and the supporting math in order to do this. In writing this book, I have done quite a bit of the same. I have researched formulas for the power requirements of cryogenic chillers; I have studied radiation impacts, and so many other topics. My background is structures, strength of materials, aerodynamics, and rocket propulsion.

I have also covered numerous other topics in this book, everything from farming to compressors. I do not claim to be an authority on every topic but I would say I have learned enough to make valid statements based on research and to do the math on these topics. We can go to Mars, and I believe we will.

Why

I don't think the human race will survive the next thousand years, unless we spread into space. There are too many accidents that can befall life on a single planet. But I am an optimist. We will reach out to the stars.

Stephen Hawking, astrophysicist

Life on Earth may be unique. We don't know. There may be many planets with intelligent life but there also might not be. What if we are the only intelligent life in our galaxy or even just the only intelligent life for a hundred light years (600,000,000,000,000 miles)? That would make our existence very precious, but even if we aren't the only intelligence, I think the human species is still very, very precious. Wouldn't we do almost anything to save our entire species? I love my children and want them to survive, more than life itself. I also love my grandchildren and want them to survive. Why not 20 generations down the line even if I never meet them? I for one think it is imperative that our species survives. I believe life is precious and especially sentient life. "I think and therefore I am," as Descartes said. He used it as a philosophical proof for something else, but I think it also could just mean 'I matter'. I believe that life, and especially intelligence and self-awareness, matter a lot. I want us to persist no matter what happens. I hope you do too.

That is why we must do something. This Earth is vulnerable.

Meteors have struck the Earth before. It is now an accepted fact that one of them wiped out the dinosaurs sixty-million years ago. The crater scientists believe is the impact that caused it is off the coast of the Yucatan Peninsula in the Gulf of Mexico. We have craters to show for others too. The extent of damage is caused by the size and makeup of the meteor. One struck Russia in 1908, referred to as the Tunguska event (google it). It was apparently a rocky object, only about 50–60 meters across. It didn't even make it to the ground. It exploded in the air but blew over an estimated 8,000,000 trees across 2150 km^2

(830 miles^2)[1]. That is a radius of 26 kilometers (16 miles), larger than many of our cities. The destruction is shown in the photograph in Figure 1.

The Tunguska asteroid may have been too small to even draw attention with current tracking but could have wiped out a city. Its detonation exceeded the power of almost any nuclear weapon exploded to date just because of its speed coming in (about 97,000 km/hour). However, some objects flying around in space are mostly iron and much bigger. It is estimated that to wipe out all life on Earth (burn us alive, so you kill bacteria too) would take an object about 96 km (60 miles) in diameter [2]. This is NASA's number and why they have an 'asteroid watch'.

However, even one 11 or 12 km in diameter, similar to what wiped out the dinosaurs, would throw up a dust cloud that would cover the earth for years killing most vegetation, most animals, and probably billions of humans. Some would survive but that is probably worse than any disaster movie ever. Who knows what, if anything, would be left of our civilization.

Figure 1: Tunguska Event in 1908 flattened 2,150 square kilometers of forest. It was only a 50–60 meter wide meteorite.

There are many other ways we could be destroyed. What if someone started a nuclear war? It is scary how easily this could happen. We have multiple super powers and several lesser powers that have nuclear weapons. The leaders of some of these countries change frequently with half the population often thinking the wrong person or possibly even a dangerous person is in power. (Maybe one of them will be.) Other countries are more autocratic. Their leaders rarely change and may continue in power after they have lost their rational mental capabilities. Many westerners believe that is the case already in at least one country. There are other countries that if they get a nuclear weapon might use it. This could easily lead to escalation.

Assuming it is all-out between two super-powers, it would probably be in the class of the wipe-out-the-dinosaurs event described above. The detonation of thousands of nuclear weapons would poison much of the planet for decades and possibly create dust clouds resulting in a 'nuclear winter'. That is close to a species killer.

Another issue is the planet's natural resources. We rely on them. We need them for the power to run industry, to heat our houses, and to get to work. Eventually, we will use them up. Why not do something about that while we still have resources and can. And don't just say we will use 'renewable resources'. Solar power requires silicon and other materials. The silicon manufacturing process is one of the major sources of fluorine-based greenhouse gases, some of which have hundreds or even thousands of times the infrared absorption of carbon dioxide. I am not saying no solar at all. I am all for solar. I am just saying that it has its limits too. Eventually, we will have no resources left or will have ruined the planet. We have the resources now.

Many people think we are killing ourselves with global warming. I am not an alarmist on this at all but will say we are having some effect and it could get a lot worse if we don't make the right decisions. I don't mean in the next few years but certainly over time. We are putting some greenhouse gases in the atmosphere that have lives of thousands of years and absorb infrared (the cause of a greenhouse effect) at hundreds or even thousands of times the rate of CO2. The quantities are not huge on a planetary scale, but if we keep doing it for decades or even centuries we will have an irreversible effect that could last for thousands of years. Also, how do we control what the worst polluters like China and India do? They are exempt from the Paris Accord as 'developing nations' (who negotiated that with the number 2 and number 6 economies in the world).

One final thing that WILL destroy the Earth is the sun. It is growing and expanding and eventually will turn the Earth into a scorching lifeless desert. This is not going to happen for millions of years, but the point is the Earth will eventually be uninhabitable. The only question is when. If we do it, it will probably happen soon. If an asteroid or comet does it, it could happen any time, though a planet destroying event is obviously not that frequent. However, our species is too precious to leave our existence to chance.

So what do we do about this, while we can? A very good solution is for our species to move to more than one world. That protects us from most of the possible ends to the species at least.

So how do we do this? Let's start with where we would go.

We start by looking in our own solar system. There are many planets and moons, but most are uninhabitable. They lack the resources; they lack the conditions, including our own moon. It has no air, very little water, and has burning hot days and freezing nights. Venus and Mercury are hotter and worse. Jupiter and out from there have a tiny fraction of our sunlight and cryogenic temperatures.

Where do we look then? The best answer is Mars. Is it Earth-like? No. Can it sustain life as it is? Well, maybe no, but it does have the things we need to survive there using a little bit of engineering. It has a CO2 atmosphere. It has water, actually a lot of it, and we can use that for consumption, making oxygen, and watering crops. The combination of CO2 and water mean we can make rocket fuel, which gives us a way to go back and forth. Finally, we could take inflatable greenhouses and grow our own food. So yes, it has what we need to survive if we apply our skills at harnessing its resources. This is not true of anywhere else in the solar system. It is also one of our closest neighbors. Therefore, Mars answers the 'where'.

We are in a deep gravity well on Earth and it is hard to go anywhere else. That (plus a lot of bureaucracy) is why governments have spent billions to get humans anywhere, even just to orbit. However, getting to orbit can be done at much lower and steadily falling costs by private industry and if we can produce rocket fuel on Mars it gets a lot easier to go there and back and we can. Robert Zubrin made this point in his book, *The Case for Mars*. Being able to make rocket fuel there is critical to credible, cost-effective travel.

So now we know where and just a thread of how. Join me for a much deeper journey on how we will do it. As Plato once said: "I will not say this is a true story, but I will say that it is a likely story." By that I mean I am not a SpaceX employee nor do I have access to their detailed planning. In fact, they may not have planned this far, but I will tell you how it can and perhaps will be done. I started this journey by confirming their math and wound up with a potential architecture for colonization.

History

Our two greatest problems are gravity and paperwork. We can lick gravity but sometimes the paperwork is overwhelming.

Werner Von Braun, rocket scientist

The Early Days

The concept of a manned mission to Mars was created by Werner Von Braun. Many know him as the father of the Apollo Program. He was also the chief architect of the V2 rocket used by Germany during World War II to bomb England. At the end of World War II we snuck him and many other German engineers and scientists out of Germany and brought them to America. He worked here first on ballistic missiles for the Army and later for NASA.

Way back in the late 1940's after being brought to America at the end of World War II, he conceived the idea of a manned mission to Mars and worked out the details. He wrote a book on it in 1949 in German. It was later translated to English and the English version was rereleased in 1952. It was called ***The Mars Project*** (or ***Das Marsprojekt***). The cover is shown in Figure 2. It described a fleet of 10 rockets, each with a mass of 3,720 metric tons, which are assembled in orbit at a space station. Three are unmanned and transport a winged landing craft each, as shown on the cover. The other seven carry a total of 70 astronauts. He clearly had done his homework. All of the math was described and it is reported to have been accurate. He was brilliant and definitely thought big.

Figure 2: The Mars Project, Werner Von Braun's 1949 book

The book described using Hohmann transfer orbits to get to Mars and back, which is the least possible energy trajectory from one orbit (Earth around the Sun) to another (Mars around the Sun). He even worked out the supplies needed for the crew. Several years later, perhaps when he realized the cost involved and federal budgets, he created a scaled down concept. It only involved 2 ships, one with astronauts and one with cargo.

He later became the director of the Marshall Space Flight center and the chief architect of the Saturn V rocket that took us to the moon. (I am sure NASA drove him to the quote at the start of this chapter).

In the 60s, when we were on our way to the moon, everyone thought we would go to Mars by 1980. It was inevitable. Von Braun had designed the Nova rocket to get us there (see Figure 3). We just had to build it. However, Richard Nixon was not terribly interested. When we landed on the moon in July of 1969, the government contractors laid off 50% of their workers. I was told it was on the day of the landing. I am not sure that is true or that I could verify it even if it was, but the point is we cut the hell out of the program.

We had done it. It was over.

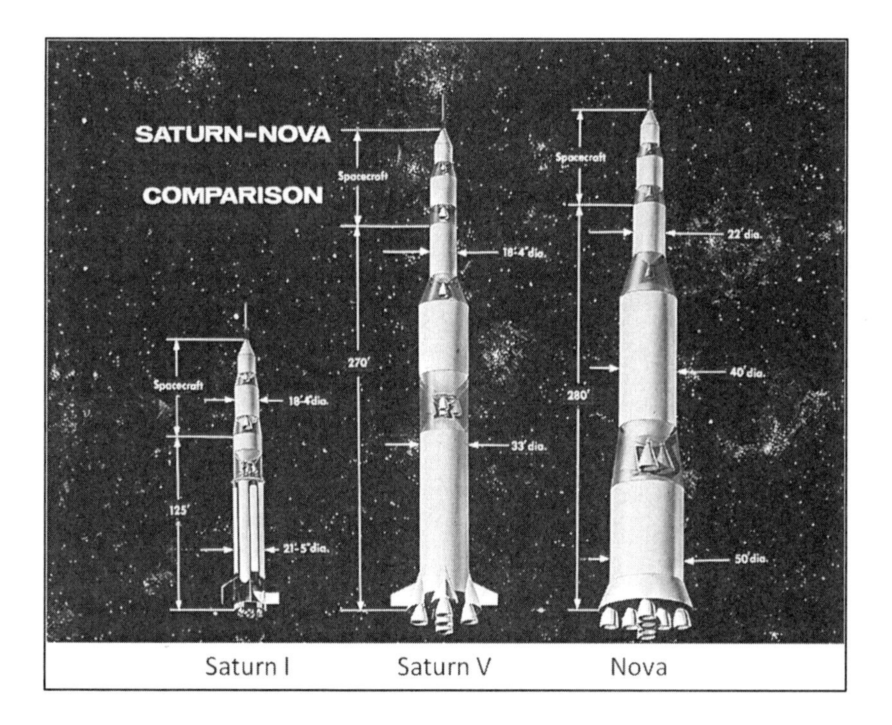

Figure 3: The Nova rocket that was going to take us to Mars

Well, not quite over. We actually launched 7 missions to the moon. Six made it there, the exception being Apollo 13, so we did continue but the budget went way down. It ended after Apollo 17 in December of 1972, a month after Nixon's re-election. There was supposed to be an Apollo 18 but it was cancelled.

We had a huge program to get us to the moon because John Kennedy issued that challenge on 12 September 1962: "We choose to go to the moon in this decade and do the other things, not because they are easy, but because they are hard." However, once we planted our flag there 6 times it was over. Obviously, they did a lot of science too, but to spend all of that time and money to spend two to five days on the moon was not the kind of scientific expedition I am proposing in this book. We need to go to Mars and stay there.

After the Apollo program, we started doing things like Apollo-Soyuz where it was more political than science and also Spacelab. The new initiative was the Space Shuttle. It was low-Earth orbit (LEO) and not true exploration. I say that having worked on the Space Shuttle program most of my career. It was an incredible vehicle, but it really wasn't exploration. It was more like 'let's do experiments in LEO forever'. There has been some value in those experiments and certainly in payloads like the Hubble Telescope, but it is time to move on.

The shuttle flew from 1981 to 2011. The SEI description that follows was during this period.

Space Exploration Initiative (SEI)

In 1989, President George Bush picked up the vision. He wanted to send people to Mars. He commissioned NASA to come back with a proposal. He wanted a long-term vision, not a 2 year budget plan but long term goals as to how we might go to the moon and Mars. This was bold and it was far-sighted, but you have to be careful what you ask for from a government agency. It is also unfortunate that he said the moon and Mars. We had been to the moon and going there is not a requirement for going to Mars, in any way.

NASA brought together every NASA center to work on this for a period of three months. It became known as 'The 90 Day Project'. Anyone familiar with federal budgets and government agencies knows the result of that. Every department and every center came back with their 'crucial' role in the initiative and their budget for it.

Do you think any department or any center said, "Nah, we aren't necessary"? Of course not. We had to study the effects of weightlessness or build a base on the moon or do something, often unrelated to the end goal but they put it in a context where it was vital.

I think they believed that after the Apollo program had gotten so much funding anything was possible. They were also fighting for their jobs in every department and at every site. Perhaps George Bush would make a 'We will go to the moon' kind of speech and this can all get funded.

The proposal that came back was a monstrosity. It said we needed a 30-year build-up before we could go to Mars. It involved bases on the moon (why?), assembly on orbit, and basically funded every pet project imaginable. The space station had to double or triple in size so we could have huge docks to build and service ships. We had to have stand-alone fueling stations in orbit with cryogenic propellants. The infrastructure was incredible. The price tag came back at $450B. That was almost half of the total annual federal budget at that time. It might be spread over several years but there was no way NASA was getting even a couple of percent of the total national budget, especially when it is connected to a total price tag like that. (NASA was getting less than 1% of the federal budget during the Shuttle program, even if it was billions a year, except for a brief spike just over 1% during the Bush administration.) I won't say it was laughed out of congress, but it got very few votes. Every time an initiative came before congress with funding for SEI, it was removed or zeroed out. It was the most extravagant, most expensive program that could ever be imagined.

Whoever was integrating the study failed miserably. They should have fought back on the absurd inputs and redlined things out. We could have gone to Mars for a fraction of what finally came out in the report and we could have done it in ten years without space stations or moon bases.

SEI never made it anywhere and NASA's budget remained a fraction of SEI levels.

Zubrin and a Way to Go

In 1996, Robert Zubrin published a book *The Case for Mars*. He used the term 'Mars Direct' as in go straight there without any space stations or moon bases. It laid out a logical plan that would get us to Mars multiple times in the $30B range. It used in situ (on site) propellant production on Mars. The idea was that we could save a LOT of mass if we could produce propellants on Mars for our return trip. It made sense. It works. In fact, his book is what started me on this path though I didn't read it until 2022.

In it, he described the idea of sending your return craft to Mars ahead of you. It has the fuel factory and refuels itself. The astronauts never leave Earth until they know their return craft is fueled and ready to go. They land near it, live on their own craft on the surface for a year and a half doing science until the return window opens. Then they move to the return vehicle and travel back to Earth.

The propellant production is the same concept I will describe later in this book. He didn't actually offer any schematics of the system or a lot of the details. However, he demonstrated it by building the propellant production capability while at Lockheed Martin and described the concepts. This production of methane and oxygen on the surface of Mars from materials there is a brilliant step forward that makes it possible to go a lot smaller and a lot cheaper to Mars and back. These are the fuel and oxidizer that will power the rocket engines to get us back to Earth.

The cost was far lower and the ships far smaller because we don't need to take all of the fuel to get back with us. However, he was still throwing rockets away, which is why the cost was still in the tens of billions.

NASA has looked at this idea and has considered it. Whether they ever try to do it or not is not clear, but I was recruited by a team at Kennedy Space center back in about 2003 to model this fuel production system mathematically. I did, though I do not still have the model. It was in a software tool I am not sure still exists and I didn't keep copies due to corporate ownership policies. (I wish I did have it). I guess I could recreate it but haven't. I just researched the equipment and its requirements for this book.

Anyway, the ideas in *The Case for Mars* were brilliant, the narrative logical, and the book compelling. As I said, it helped start me down this journey. As I got started, I stumbled across SpaceX's 'Starship' and an Elon Musk update. I was not aware at the time that SpaceX had anything this far along, but then again, I don't watch a lot of news and I am not all over social media.

If you have read Robert Zubrin's book, you will notice that while I go the same route on a few things, I go on a very different path on others. Part of this is the reality that SpaceX will do this and how. Part of it is a disagreement on a few topics, partly just because of new information. For example, I do not believe taking hydrogen along is an option, at least not on SpaceX's Starship, as I will explain later. We have to find water. Then again, back in 1996 we didn't know if there was water on Mars. We now know there is water and lots of it.

I met Robert Zubrin at a recent conference. I spoke with him and listened to him present. I was quite impressed with his logic and the things he said. One thing he said was that space can only be explored and exploited by a free society. China or Russia would never allow or provide the environment for the creation of an Elon Musk or a Jeff Bezos, who both have their own space program, and new ideas are never created in a tyrannical society. It is our freedom that allows us to dream and to create. I think I believe that. After all, name all of the inventions of the last two centuries. They all happened in free countries, America somewhat dominant among them but certainly not the only one. America is rich because it is free. We do as we please and we make our living as we please (within some legal bounds preventing harm to others). People start companies and market many creative ideas. A bureaucracy just can't do that.

SpaceX and the Vision

Elon Musk became a billionaire early on in about 2002 at the age of 31 and started SpaceX. His goal as stated was to colonize Mars. He started doing this by making trips to orbit quite a bit cheaper, which is a first step in facilitating Mars. It is reported that he read Robert Zubrin's book and was inspired by it. I don't know if he has confirmed that but he certainly has a passion for Mars and says he is using the in situ propellant production concept (and the latest edition of Robert Zubrin's book has a photograph of the two of them together).

Elon Musk earned Physics and Finance degrees from the University of Pennsylvania before starting SpaceX, but dropped out of a graduate program to pursue his business interests. He was the founder of PayPal and SpaceX among others, and long-time CEO of Tesla, so his business instincts and his applied physics are both excellent, though he hired a lot of talent to help him with the rockets.

His company SpaceX created the Falcon 9 rocket, with the first truly reusable booster. It has been used to launch satellites and to take crew and cargo to the Space Station. It has revolutionized launch to orbit. Because it is mostly reusable, they can get a payload to orbit for about half of what other companies can. As a result, SpaceX is the number one private launch company in the world, in terms of number of launches.

Now, SpaceX has come up with the Starship launch vehicle. It is 100% reusable and can perhaps drop launch costs by another order of magnitude. As he says, you have to run space travel like an airline if you want to get costs down. You have to be able to launch again quickly like the airlines do. If you do, the cost comes down to almost just the cost of fuel (his statement, not mine).

Starship—A Whole New Game

A foolish person has a little money and throws it away on things they don't need. A wise person has a little money and invests it. That makes all the difference. Years later, the wise person has everything they hoped for. The foolish person has nothing to show for it.

Anonymous

As of October 2022, the human race has never launched a spaceship. Absolutely never!

Let me explain.

In the last half a dozen decades, we have only launched artillery. What I mean by that is a spaceship is reusable. Rockets based on ICBMs are artillery. They throw away pieces along the way in order to deliver their projectile and are expended. That is the way artillery works and it is not sustainable. If you fire a bullet and somehow catch it, that just means you didn't shoot anyone. It is still artillery. The same is true of capsules. An ocean-going ship comes back to port. It doesn't drop pieces along the way. A true spaceship is like that. It all comes back to be used again and again. You can't create a navy on artillery. You need ships you can use over and over.

A gentleman at a conference I went to recently, Bill Bruner, made this point. He put pictures of Starship and NASA's new rocket Artemis side by side. He said we compare rockets in many ways, such as thrust or height or payload. However, these two are about the same size but nothing at all alike. One is a spaceship. The other is artillery. I suppose that is true. Artemis will be the most expensive kilogram to orbit vehicle ever at somewhere over $44,000 per kilogram and no piece of it is reusable. When you drop four $146M (reusable) space shuttle engines in the ocean, along with a lot of other hardware, it gets really expensive.

On the other hand, Starship will be the cheapest ever by a factor of 4 over the cheapest yet, a SpaceX Falcon 9 Heavy, at about $400 per kilogram initially (my estimate not their number assuming $50M per launch initially) and that could drop by almost another order of magnitude by the fiftieth or hundredth launch to somewhere around $80 per kilogram (my estimate at $10M per launch but more than their number). That would be about 550 times cheaper than the Artemis cost per kilogram to orbit.

For decades, we have been throwing away millions of dollars of rocket hardware on every launch. This was just the way they did it because it was easy and efficient as far as a simple design, while ignoring cost. I won't say Elon Musk was the first one to consider reusability. However, it appears he and the engineers at SpaceX have finally done it.

There were a few cases that were better than zero reusability. The Space Shuttle brought the orbiter back and reused it. However, the processing between missions took months, the costs were outrageous and we were still throwing away the External Tank every time, by dropping whatever of it didn't burn up into the Indian Ocean. The cost for a Shuttle Launch came in somewhere over $1B per launch, in 2022 dollars. This is based on a total program cost of $196B in today's dollars spread over 135 launches.

I don't consider the Shuttle SRBs to even count as reusable. They were torn down and completely overhauled every mission. The cost was about the same as building it in the first place. Because of this the new NASA rocket Artemis will throw them away rather than try to reuse them.

Commercial companies started launching payloads several years ago. They were a little expensive, but definitely cheaper than anything the federal government could ever do. The cost for an Atlas Launch, for example is currently about $110M–$153M depending on the payload mass and therefore the rocket configuration. SpaceX came in as a competitor several years ago with the Falcon 9 at around $50M and slowly started taking business away as customers gained confidence in their ability to launch successfully. Due to inflation, that price is now up to $67M for the Falcon 9 or $97M for a Falcon Heavy. It is cheaper partly due to efficiencies, but also very much due to flying the same boosters over and over. However, the upper stage is not reusable.

Now Starship comes along. First of all it is completely reusable. It is also huge. The 'Super Heavy' booster is 70M (230 ft) tall and the upper stage is 50M (165 ft), for a total height of 120M (394 ft). The vehicle is 9 meters (29.5

ft) in diameter, except for the nose, which tapers in for aerodynamics. The entire vehicle has a mass of 'right around' 5,000 tons (11M lbs.) fully fueled according to Elon Musk. My best estimate right now is 5,090 tons (11.2M lbs.) including payload or 4,970 without, but as I said it is an estimate and they are still trying to shave weight off. The current dry weight I am using for all calculations is 370 tons total. That is both stages and does not include any payload. The reason I am using that number is it is the last number Elon Musk and SpaceX have given us. They are trying to drop weight but this number will suffice until they let us know that they did.

Super Heavy was 250 tons (551,000 pounds) according to Elon Musk in the spring of 2022. He says they will get that down to 200 tons, but until they announce a new mass, I will use 250 tons. This mass still gets us to orbit, though a little lighter would increase payload a bit. That is 6.8% of the total mass for the engines and structure. That is actually very good to be 93.2% fuel by mass. We will talk about why in the next chapter. The Super Heavy has 33 Raptor 2 engines at 1,600 kg each (someone else's estimate based on a statement from SpaceX of being 20% lighter than the Raptor 1 which was believed to be 2 tons). The engines are a total mass of about 53 tons, so that leaves 197 tons for structure. (Note I will use the term 'tons' throughout this book. It refers to a metric ton, which is 1,000kg or 2,205 pounds.)

The Super Heavy is capable of holding somewhere between 3400 and 3,600 tons of fuel. I am going to assume the worst case of 3,400 tons (7.50 M lbs.), but only because I have heard both and want to err on the side of being conservative. (Conservative in this case is an engineering term for be cautious rather than assuming best case). I believe it is also the more widely accepted number. With super-chilled propellants, which they have said they are using, the density goes up about 10% and 3,600 tons might be possible, though it depends on the dome shape on the ends of the tanks and how much space between the engines and tank. Either works to get to orbit, but as I said with the dry weight, the other is better. 3,600 tons would raise payload just a little but I am going to assume 3,400 tons.

The Super Heavy shuts down and separates after using about 92–94% of its fuel. This early shutdown is so it can fly back to the launch pad where it lifted off. I used 92.6% in calculations because I did not do the entire calculation on the (sort of) aerodynamic entry back into the atmosphere, but

7.4% (250 tons) is enough to reverse its velocity and head back in a ballistic trajectory with some fuel left to slow down at the end so I am very close.

The intent, as Super Heavy is coming down, huge arms on the launch pad, often referred to as 'the chopsticks' will catch the Super Heavy. They actually squeeze in while it is still firing and moving about 2–3 meters (6–10 ft) per second but catch on lugs located near the top of Super Heavy. Once they catch it, they rotate it around, center it, and set it down gently in the correct spot to launch again.

This never before done feat is reported to save about 5–10 tons of structure that would be needed on the booster to protect it from a harder landing. Even Elon Musk says this is a hoped for outcome, not a given. That may be true but he has never been afraid to fail and then tweak the design based on what they learned. It happened like that with the Falcon 1, the Falcon 9 booster landing, and the first Starship attempts at an aero entry with a vertical landing in a test flight. In every case they got it right after a couple of attempts, so I am betting they will in this case, even if they lose a booster trying the first time.

The upper stage, referred to as Starship has a mass of 120 tons according to an update from Elon Musk about a year ago and according to their website currently. He said he thought there was a path to 110 tons but what he really wanted was 99 tons. That may be wishful thinking. Some of the early versions failed on landing due to structural issues before they modified the design and the software for landing. I will go with 120 tons until I hear otherwise from SpaceX.

The Starship has 6 Raptor engines. The 3 inner engines are sea-level (shorter nozzle) and would be the only engines used to land, especially on Earth. They can gimbal to direct the thrust for steering. The 3 outer engines have longer nozzles and are optimized for operation in a vacuum. They are more efficient, but would probably not be used at sea level. They do not gimbal.

Optimizing for a vacuum means a longer bell. As the exhaust expands out of the bell shaped nozzle it accelerates the exhaust faster and faster to very high speeds, as in 13,500 kph (8,387 mph) in the case of the vacuum engine. The faster it goes the more thrust you get and the more efficient the engine. However, as it is doing this the pressure drops lower and lower. If it dropped low enough it is possible for the bell to actually collapse in the atmosphere depending on how strong it is. However, just the fact that you expand the exhaust below atmospheric pressure means there are shocks in the plume and you don't get full efficiency. Even the sea level engines expand it a little below sea level pressure apparently because they get an Isp (efficiency) of 327 at sea level but about 355 in a vacuum. The vacuum optimized engines have a longer, wider bell and achieve an Isp of about 382 in a vacuum.

Isp is the thrust divided by mass flow and has units of seconds. Think of it as how much thrust can you get for a given amount of fuel. More thrust for the same amount of fuel is better obviously, but it happens in space or at high altitudes, more than at sea level. By this I just mean maximum efficiency. The Raptor engines are pretty efficient even at sea level. However, they are more efficient in space, like most rocket engines are.

One other efficiency is the density of methane. NASA has often used hydrogen, but it has such a low density that the resulting rocket winds up being huge and requires a very large, heavy tank. Two examples of this were the external tank on the Space Shuttle and the first stage on the new Artemis rocket. (Liquid hydrogen is also very expensive).

Super-chilled methane is over 6 times as dense as liquid hydrogen. This results in a much smaller volume for the rocket, and a lot less structure. For this reason most commercial launchers use propellants other than liquid hydrogen. Most use RP1 (rocket grade kerosene) and liquid oxygen, but liquid methane has higher performance and is also the propellant we can readily produce on Mars.

Base Vehicle	**Mass (tonnes)**
Booster Fuel	3400
Booster Dry Mass	250
Booster Total	3650
US Propellant	1200
US Dry Mass	120
US Starting Mass	1320
Stack Mass	4970
Payload	120
Stack Mass w/ Payload	5090

Table 1: Starship Mass Summary

Starship (upper stage), which uses methane and oxygen also, can carry 1,200 tons (2.65M lbs.) of fuel according to SpaceX. It also can be refueled on orbit by tankers, which would be needed for trips like the Moon or Mars because it would not have enough fuel left for the trip once it reaches orbit. The refueling in orbit and on Mars is why it can do the round trip at all, and yes, the intent is to land this huge Starship (upper stage) on Mars. They advertise 100+ tons to orbit, the Moon, or Mars. That is a little ambiguous but it can certainly get 120 tons to low Earth orbit so I am going to use that number. If it can only land on Mars with 100 tons, it does not change the premise of this book, only the mission-planning details.

I believe their reasoning for being ambiguous is that every time they modify the rocket, the actual number goes up or down a little. They don't want to advertise a different number every couple of months. They don't know the final number until they finalize the design. Also, the payload to orbit varies with orbital altitude and inclination. A higher orbit means less payload. Therefore, they may always just say 100+ and provide an exact number when a client gives them a desired mission.

These masses are summarized in Table 1. These are the masses used in calculations for launch. For the trip to Mars and the return it is the upper stage but with only the propellant needed, not full tanks. Landing with a large amount of excess propellant would be an issue trying to decelerate, especially in Mars' thin atmosphere.

Figure 4 shows an estimate of the internal structure of Starship. The lines inside the tanks are baffles that prevent sloshing in the tank.

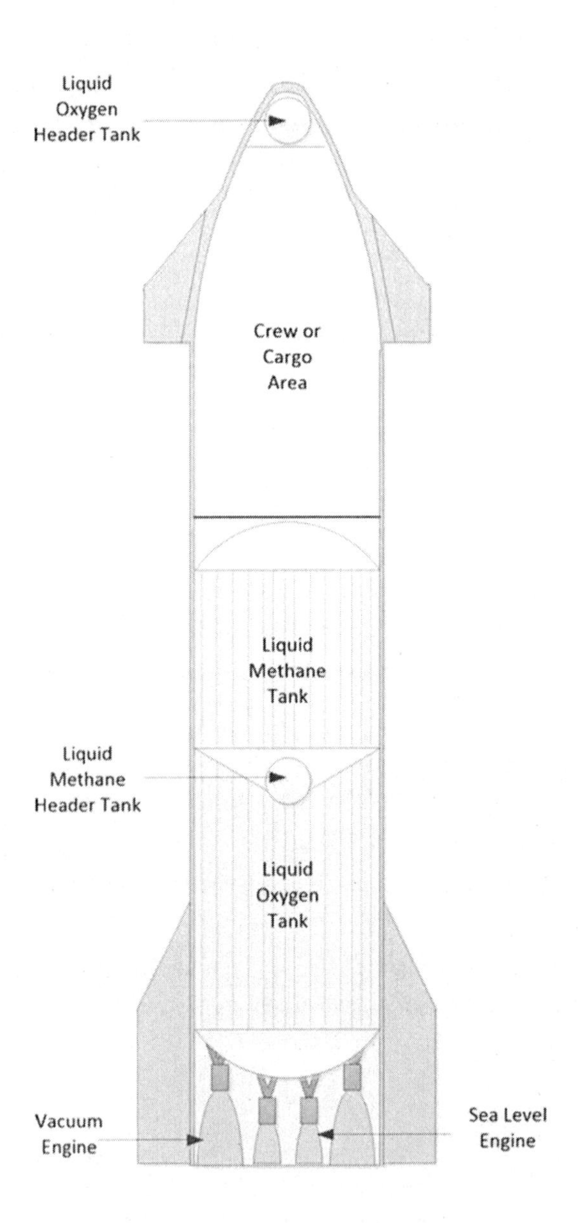

Figure 4: Cutaway of Starship

Starship flies back into the atmosphere aerodynamically, similar to the Space Shuttle Orbiter, except that when it gets near the ground it flips up and lands vertically. It can do that on Earth or on Mars, which has much less atmosphere. It has done that on Earth already, but not from orbit. SN-15 flew to about 10 kilometers, flipped over on its side and came down. When it was near the ground it did the flip upright maneuver and landed safely.

Here on Earth, it will be caught by the chopsticks as it descends, just like the booster, and presumably be set down on top of the booster to fly again. However, on the Moon or Mars, 'there is no infrastructure yet' as Elon Musk put it, so it will have legs to land on vertically just like the Falcon 9 boosters do currently.

The chopsticks rise on a rail to the right side of the vehicle from the viewer's perspective (left if you are on the tower behind it). When they get high enough to clear they rotate from right to left, again from the viewer's perspective, to place the booster on its launch mount or in this case the upper stage on the booster. I assume the Starship will be caught in front of the vertical rail, not over the booster.

The reason is you would never fire rocket exhaust at anything you care about. The Methane-Oxygen combustion temperature is somewhere around 3500 degrees C (6,300+ degrees F). It cools as it expands but if you put something in front of it you will find that it is still a very effective blow torch. The heat can be recovered by decelerating and compressing it, like a flat surface would.

I recall a test stand in White Sands, New Mexico. We had a Space Shuttle OMS Pod that was used for testing system performance back in the 1980's. (Those were the two bumps on the back end). The OMS Pod rolled out of the hanger on a rail to fire the engines. However, it didn't roll out quite far enough for the last thruster to fire without hitting the building. The test engineer wanted to fire it, so he decided to create a deflector. He put a ¼" stainless steel plate in front of it at a 45-degree angle to deflect the flame and protect the building. Note that a Shuttle RCS thruster is a smaller engine, about 0.4 tons (870 lbs.) of thrust versus 230 tons (510,000 lbs.) of thrust for a Raptor engine. It is about as violent but on a much smaller scale. The exit is roughly 0.2 meters where a Raptor exit is 1.3 (sea level) – 2.8 (vacuum) meters wide.

When the test engineer fired the thruster, within seconds the team saw molten steel running down in a stream on one of the cameras. It melted a hole right through the stainless steel plate.

As an ex-rocket propulsion engineer, it seems to me that firing the engines at the booster as the upper stage descends has a lot of potential to damage the booster beyond repair. I am sure they are very aware of that too. Instead they will catch in on the vertical lift path it goes through for stacking (or even farther away if the chopsticks rotate farther). The chopsticks will then lift it to the right level, rotate it over the booster and lower it to mate it. The booster is probably caught this way also, rather than over the launch mount.

When the Starship is lifted to a point higher than Super Heavy, the chopsticks will rotate to its right, our left. After it is lined up, it will lower Starship and mate it to Super Heavy. They have not caught Starship as of this writing but the chopsticks have been used for stacking. The chopsticks lifted Starship from its transport vehicle on the ground.

Note that since the booster comes back to the pad within minutes after launch, whereas the Starship could be on a mission for days or even years (Mars) the next stack will often be that booster with a different Starship. This implies a need for interchangeability and also a need for a lot more Starships than boosters once they are in full operations and doing missions like Mars trips.

Once the two are stacked, payload and/or crew have to be loaded, and they have to fuel it before launch. Elon has said it could theoretically launch again in an hour. I think he means the rocket is ready to go again not that it will actually launch that quickly.

A payload loading time under several hours would be amazing, and even that only achievable with a payload that they are willing to already have at the pad while they are catching rockets or a payload rolling in from nearby for a crane to lift it right after the Starship is mated to the booster. Fueling also will take several hours after the payload is onboard and sealed up. Assuming monstrous pumps that can load 30 tons/minute, it would still take almost 3 hours, plus whatever startup and throttle down on the filling operation. So, depending on the payload a turn-around of a day to a couple of days is certainly achievable with around the clock operations.

However, having said that it isn't 'super quick' like an airline, this is still light-years ahead of anything we have now. There is no discarded hardware and a single Starship might be able to launch more than once a week, possibly even daily, depending on its mission.

If you can reuse the vehicle over and over, the cost comes down to fuel, labor, any maintenance, and prorating the cost of the vehicle over its useful life. The fuel portion currently is about $900K according to Elon Musk. However, he has announced they are going to start making their own. This eliminates transportation and tanker costs and probably reduces fuel cost by 50–90%. Bringing in 4,600 tons of propellants requires at least 170 cryogenic tankers traveling from the supplier's plant to the pad, and pushes the cost way up.

Elon Musk says he is going to remove CO2 from the air to make rocket fuel. If so, the only other thing he needs is water using the process that will be described in chapter 6. Removing CO2 from the air by condensing it out is nearly a futile task because our air is only 0.042% CO2. In other words, you would have to compress and chill 1,000 kg of air to about -50 ºC to condense out 0.42kg (1 lb.) of CO2. This process would include condensing out about 25–30 kg of water in Florida (probably similar at the other pad in coastal south Texas too). I have that cost at about $20K per ton just for electricity.

For this reason, I believe he is going to use direct capture, a filtration system to remove CO2 from air. It involves filters that capture the CO2. Once the filter is saturated, you put it in an oven and bake it in order to release the CO2. It is currently reported as under $100 per ton and perhaps falling. You need about 57% of the total mass of propellants of CO2. That is about $200K to $250K per launch, but maybe lower soon or lower at the incredible scale they would be at for even 1 launch per week (2,736 tons CO2 per week). The CO2 is the hardest part. The rest of the process is probably about $75K in electricity per launch (4,600 tons) at Florida's current rate of $0.14 per kWH, which is four cents below the national average.

The cost of a launch is quite a bit more than just the fuel however. I won't say he can do it for $1m a launch, but I bet the cost slowly drops to the low millions per launch as they work it up to a couple of launches a week or more. And that is for 100+ tons to orbit, far more payload than competing rockets. In fact, things like a tanker launch where the only payload is filling extra tanks should be one a day kind of rate or close to it, with one booster and two or three tankers. The tankers have to transfer 100+ tons of propellant to the vehicle on orbit. I have no idea of the transfer rate. If it is fast enough they could perhaps launch 2 tankers in a day.

SpaceX has at least one contract for a payload in 2024, two lunar orbit missions for paying private individuals, and NASA announced that they will pay SpaceX $4.2B to develop Starship for two NASA moon landings, so there is definitely a market for Starship, probably a lot more after a successful launch or two. Whether there is a market for a couple of commercial launches a week is in question. However, as prices come down the market grows. If you had a restaurant and sold meals for $20 each, you might get 150 customers a day in a mid-sized town, assuming the food is good. If you are able to sell them for $2, you would have a line out the door all day long and sell thousands of meals per day. It is the same concept. As the price gets lower, it makes sense to a lot more people or a lot more companies to send payloads into space.

There has been talk for decades about things like space hotels or space tourism. If it gets cheap enough it might be surprising what kind of business models work. There are already a couple of billionaires that have paid their own way to go to space. They paid tens of millions of dollars. What if the price were $100,000 to stay on orbit for a week? I believe there are thousands of people or more that could afford that and would go for this 'once in a lifetime' experience. There are businesses now built around renting yachts for tens of thousands of dollars a week with a full crew, just like the TV show *Below Deck*. Why not a visit to space instead if you have that kind of money? You may not be catered to, but the view is incredible and how many people can say they have been to space?

SpaceX also wants to use missions for their Starlink satellites and the colonization of Mars. That will be dozens or even hundreds of flights, but hopefully they will have a lot more business. SpaceX colonizing Mars is a wonderful thing for humanity. Hopefully, they can pay for it along the way by launching to orbit for pennies on the dollar of other launch companies and do even more business with NASA.

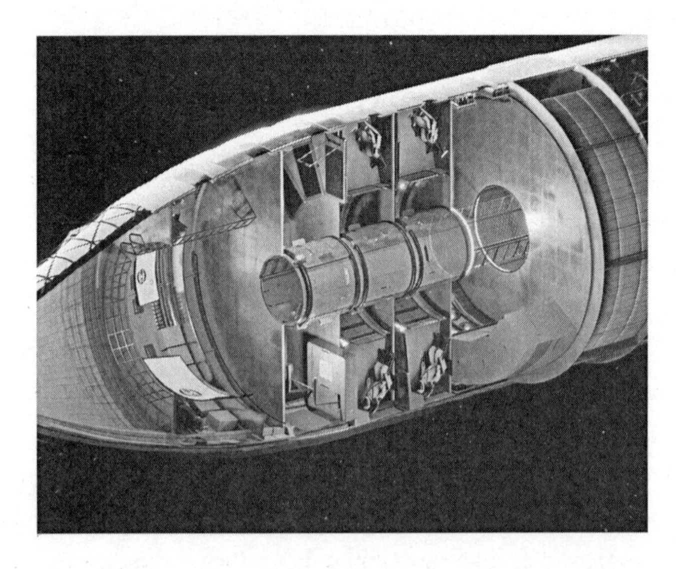

Figure 5: A concept of what starship might look like inside on a mostly crew mission (artist's concept)

Screenshot from a YouTube video. Linked from humanmars.net. Credited to 'two brothers from Deep Space Courier'.

SpaceX will use many launches for their own Mars missions, but that may not occur for a few years. Once a base is established there however, even that might become commercial. What if it was an option to move to Mars permanently at a cost of $500,000 per person (Elon Musk's number), or even $1,000,000 and be a space pioneer. There are definitely people that would pay it. They would sell their house or business and buy a ticket. Even if it is only one person in one-hundred-thousand in the world that is more market than SpaceX could handle in decades. The only issue with that model is it would probably attract middle-aged people primarily that have already created some wealth. We also need younger people, especially those that will have babies in order to continue the species. How do we include them? We need to. Perhaps they get a discount or some other incentive. Perhaps every wealthy middle-aged couple has to sponsor one young person. Perhaps wealthy people take their own children with them. We don't know what their business model will be but there is a market if they can make it work.

Note that Starships are planned for crew or payload. Figure 5 shows an artist's concept of a crew version. The strictly payload version could be used for launching large (or many) satellites. In Von Braun's vision, ships are dedicated to one or the other. That makes some sense in that no life support systems are required on a payload only ship. However, a mixed model is likely for a trip to Mars, where they have crew and payload. We will discuss that more as we get into the next chapters.

As of this writing, there hasn't been a full-up launch to orbit yet due primarily to government bureaucracy. The EPA held SpaceX up for about a year in Texas. When they finally approved 5 launches per year it came with 78 mitigations, some as silly as the company must contribute to an 'Adopt an Ocelot' foundation. Others were more heavy-handed, along the line of buy everyone out and move the population away. NASA is holding them up at KSC because they fear any kind of incident would wipe out their astronaut-to-ISS capability. There is also the FAA on licensing for launches. Elon Musk has proposed to NASA building a crew access tower at pad 40 also in order to protect the human launch capability and thinks he will get through this bureaucratic mess (my words, not his). He hopes for a launch in December

2022. Perhaps it is really January but it should happen soon. NASA is stating early December.

At this moment, in early November 2022, they have the vehicle stacked and they are about to start a progressive series of tests, but will be careful along the way according to Elon Musk because a SUD, Sudden Unplanned Disassembly (meaning explosion) at the pad would set them back 6 months. They will do stacked full tanking and engine firings with more and more engines. If that all goes well, they may finally be ready for a launch this year or early next year.

I have to add one other thought. SpaceX may lose one or two before they get it right, but once they do, the sky is the limit. As Elon Musk once said, "If you don't blow something up once in a while around here you aren't innovating enough." They are trying some hard stuff that has never been tried before and sometimes you have to test it and fail (with no humans involved of course) before you get it right. It took them a couple of tries to get the Falcon 9 booster landing right and also the Starship landing right, but they did finally get it right in both cases.

I believe SpaceX going to Mars is a question of when, not if. The ship is capable. Elon Musk is motivated, and he has the funds to do it. Hopefully, other business will generate profits to fund all or most of it, but I believe he will do it either way. I only hope he gets it far enough along that path to be self-sustaining before he goes there himself.

Getting There

We choose to go to the moon in this decade and do the other things not because they are easy, but because they are hard. Because that goal will serve to organize and measure the best of our energies and skills, because that challenge is one that we're willing to accept. One we are unwilling to postpone. And therefore, as we set sail, we ask God's blessing on the most hazardous and dangerous and greatest adventure that man has ever gone.

John F Kennedy, 35th President of the United States

President Kennedy made that speech on 12 September 1962. That speech is inspiring even today. He set a very daring goal. It was in the middle of the Mercury program. Up to that point, most of the rockets the US had launched blew up. We had about four manned successes in a row by that point, but everyone was aware of all of the explosions. They were all over TV. We had just gotten to orbit in the previous year. That was a gutsy speech.

I believe we have a new challenge before us. No one has made a "We choose to go to Mars" speech or at least not as dramatically. But, if you really think about it, Elon Musk has sort of. He has stated his intent. We are going to Mars and sooner rather than later and SpaceX will do it. He also refuses to take the company public until SpaceX gets to Mars because he doesn't want it to be up to a vote. We are going.

The Rocket Formula

To go fast, you have to burn a lot of fuel. That has always been the way with rockets, mostly because you are trying to go much faster than any other form of transportation. For example, a commercial airliner goes about 800 kph (500 mph). To get to orbit, you have to go about 35 times that fast.

The rocket formula is:

delta-V (change in velocity) = Isp * g * ln (initial mass/final mass)

Don't worry about the math if that is not your thing. Let me explain it. This says you get very diminishing returns as you use more and more fuel. Isp is the efficiency of the engine expressed as thrust over mass flow (how much thrust per pound or kilogram of fuel). The units are seconds. G is the acceleration due to gravity, which converts seconds to distance per second. The natural logarithm (ln) of the mass ratio means what power you have to raise e to in order to get that ratio. E is equal to 2.718, so a mass ratio of 2.718 returns a value of 1.

Isp times g is actually your exhaust velocity. In other words the amount you speed up is equal to your exhaust velocity if you go through a mass ratio of 2.718, starting mass over ending mass because you burned that much fuel. In order to double this, your mass ratio must be e^2, or 7.38. In order to triple it your mass ratio must be e^3, or 20.1. This quickly gets out of hand. It is probably impossible currently to build a single-stage rocket with a mass ratio of 20.1 to 1, meaning only 4.7% of the mass is engines, structure, and payload.

This is why we have had 2, 3, and 4 stage rockets. You can go through a mass ratio of perhaps 2 or 3 on the first stage (remember your upper stages and payload count in your final mass), to get up to a pretty high speed. Then you drop the first stage and start all over again on the rocket equation with a smaller rocket (your second stage) that is already going several times the speed of sound. Depending on the design of the rocket, the smaller one may actually make orbital velocity, which is about 28,000 km/hr. (17,500 mph), which is 7.8 km/sec, though you started with a velocity of about 1,600 km/hr. (0.4 km/sec) due to the Earth's rotation if you launch east. The Earth's rotation is why we almost always launch eastward. It gives us a small head start of 5.7% of the speed we need.

For the methane-burning Raptor engines, Isp varies between 327 at sea level to about 382 for the three vacuum optimized engines on the upper stage in space. Just plugging 327 in as an example says a mass ratio of 2.718 buys you about 3.2 km/sec forward speed in space. This means if your empty rocket plus payload weigh 100 tons, you have to burn 171.8 tons to achieve this velocity. Since it is actually final mass, and you may have some fuel left over (hopefully if you need to land), the amount you have to burn is even higher. The speeds needed for space travel are extremely high and not even close to achievable in anything we have ever built other than a multi-stage rocket.

Getting to Orbit

Low Earth Orbit (LEO) required velocity is about 7.8 km/sec (17, 550 mph) depending on exact altitude. That is a lot of fuel to get to that speed. What is worse there are losses along the way. At launch the first 1g of thrust is just balancing gravity. Whatever you produce above that is actually accelerating you upward and up doesn't even really count. You have to go up to get out of the atmosphere, but what eventually gets you into an orbit is your horizontal velocity. That is why you see rockets tipping over sideways as they get higher and higher.

You also have the issue of aerodynamic drag until you get out of the atmosphere. That is slowing you down too.

The net result is you actually have to use enough fuel for all of the losses plus still get to 7.8 km/sec. Also, it is a minimal addition but this will usually put you in an elliptical orbit and you have to burn just a little more halfway around the planet to circularize it or you will reenter the atmosphere.

The biggest thing to minimize the losses is thrust to weight at liftoff. That is why SpaceX increased the number of engines to 33 on Starship and worked to boost the thrust even with a slight efficiency loss. It was to get to at least 1.5g at liftoff. Higher thrust means more effective thrust, because the first 1g doesn't really count. Saturn V was only 1.06g at launch, but with the expendable they chose maximum propellant over maximum efficiency. It took 30 seconds just to clear the tower. Shuttle was close to 1.2g. If it is reusable though, more acceleration is a lot more important than just more propellant. You don't want massive tanks and massive structure. The booster on Starship will range between about 1.5g and 4.25g by the time of separation assuming no throttle back at the end. The increase is because it gets lighter as it burns fuel.

Another trade-off is limiting aerodynamic losses by throttling back a little to not hit supersonic speeds until about 10 kilometers (6.2 miles) altitude. The atmosphere is thinner there and therefore super-sonic shockwaves do not drive as high of aerodynamic losses or stresses on the vehicle. This is relatively early as in the first minute or so, depending on initial thrust to weight. Throttle back probably starts perhaps around 40 seconds for Starship (my estimate, not their math) and lasts about 20–25 seconds. It occurred slightly later on Shuttle. Who can forget 'Challenger is go at throttle up' at 72 seconds. It almost makes me cry every time I hear it. It was the first launch I got to see from outside the Launch Control Room in about 2 years. I was on top of one of the buildings with several other people at work. Then it blew up. My only words a couple of minutes later were "We just killed 7 people." I walked back to the office, put my head on my desk, and cried. That was a sad day.

Anyway, the delta-V (change in velocity) is generally too high to get in a single stage. That is why rockets are 2 or 3 stages. You throw away a bunch of weight and start the mass ratio all over again. In the case of Starship, you drop a lot of weight, but it flies itself back to the launch pad. Then you start the delta-V all over again from an initial velocity of about 2.6 km/sec horizontally (including the Earth's rotation) plus some vertical velocity with the upper stage Starship.

Based on my calculations, the Starship can carry a 120 ton payload to a 250 km circular orbit and still have more than enough propellant left when it reaches low Earth orbit to return if needed. That is using a spreadsheet to perform a numerical methods approach with a 1 second interval (calculate the thrust initially, apply that for 1 second, recalculate the thrust, mass, altitude, etc. and repeat). It includes things like aerodynamic drag based on the standard atmosphere, progressively lower losses due to gravity with altitude, etc. but is not a perfect model.

This is not as exact as I could do if I spent 6 months generating a computer program and had a lot of data on drag for the Starship. I am using crude one-second intervals in a spreadsheet, but it is close, like within a couple of percent. I am estimating the vehicle masses, so I am probably plus or minus 20 tons or so on remaining propellant, but it is close and says yes, Starship can get to orbit with a pretty large payload. That is consistent with SpaceX's calculations based on what Elon Musk says and what they say on their website.

Aero Entry

I am taking this out of order just because it impacts the next section. In doing an atmospheric entry, Starship comes in very much like the Shuttle did at a high angle of attack (high angle relative to the actual forward motion, like 60 degrees nose up, especially on Mars as shown in Figure 6). This is to slow it down by maximizing drag. However, when it gets almost down, unlike Shuttle it flips up and lands on its tail standing up.

As many of us know, the Space Shuttle had thick tiles to deal with the heating. The Space Shuttle also had an aluminum frame. Starship has a stainless steel frame, which can handle much higher temperatures, about 800 degrees C (1440 degrees F) higher. As a result Starship only needs thermal protection on one side and their tiles are much thinner than Shuttle tiles.

Figure 6: What Starship might look like in an Aero-Entry in Mars' atmosphere
Credit: SpaceX

In the simulation video on their website, it claims that 99% of the energy is removed aerodynamically. However, that is not consistent with the simulation. The actual delta-V equivalent to land on Mars would be much higher than on Earth because there is so little atmosphere. The ship can shed most of its velocity but in the end has to turn its tail forward and fire engines to slow down and land. According to the simulation on their website the burn time on Mars is just over 3 times what SN 15 used to land on Earth. It also is multi-engine high thrust for the first 13 seconds, before reducing to one engine for the final approach and landing. The firing begins at Mach 2.3, which is 552 m/s on Mars. It appears to be multiple engines for 13 seconds to slow it down to about 100 m/s. This is a mass ratio of 1.13. It then probably fires one for the last 24 seconds including throttling down as it goes.

At touchdown, it would have been about 37% throttle on one engine to land softly. The weight is about 91 tons on Mars versus one engine at 246 tons. You would want about a balance with a slow downward velocity. Based on number of engines and timing, I believe this is about seven times the propellant required for an Earth landing.

The delta-V required to land on Mars is probably about 655 m/s (1,465 mph) equivalent. This means allowing a mass ratio of about 1.21 for landing, probably adding in a little bit of fuel margin on top of that. If the mass ratio required is slightly higher it makes no difference in the feasibility. It only means slightly more refueling is required on orbit. That difference is in the noise, assuming SpaceX has the data and gets the calculations right.

An Earth landing probably requires about one-seventh of this because they can use the thicker atmosphere to slow down right up until the end. For example SN 15 in its test flight only burned the engines for about 12 seconds. That was initially 3 engines to flip it but it decelerated and landed on one engine. For Earth entry, I am estimating 100 m/s and a mass ration of 1.03.

The Transfer to Mars

The plan for Starship is to refuel in orbit by flying tankers up. So how much will they need? In order to go between planets you normally do something called a Hohmann Transfer. This is an ellipse between one orbit and the other as shown in Figure 7. It is the minimum energy way of getting there. It is also slow because you are going half an orbit. In the case of Earth to Mars half an orbit to get to Mars is 259 days. That is approximately halfway between half of an Earth year and half of a Mars year.

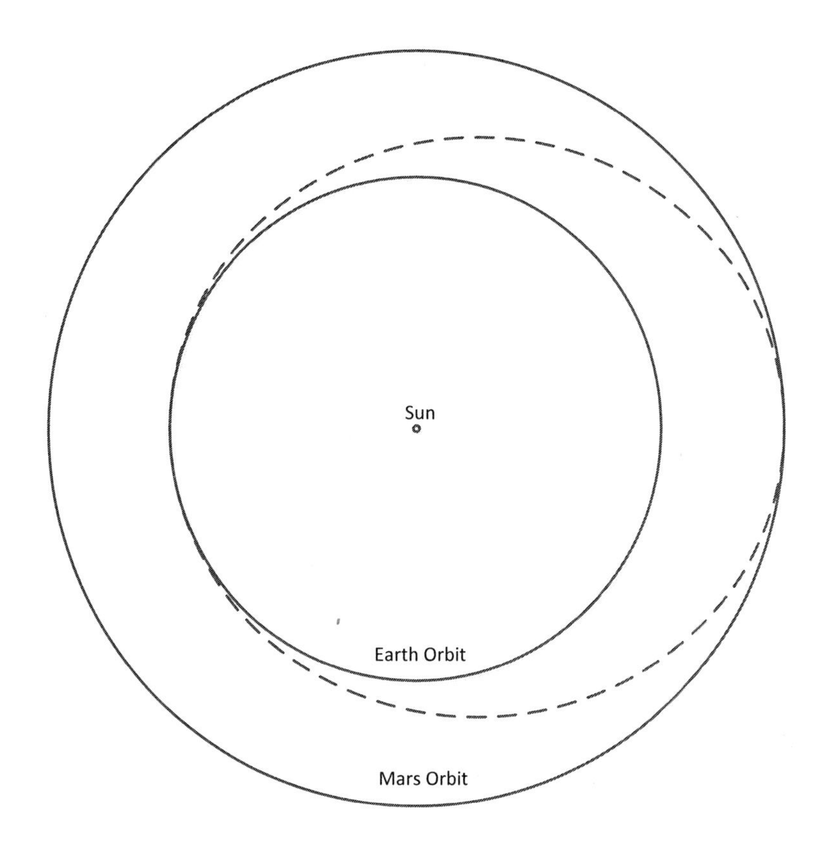

Figure 7: Hohmann Transfer Orbit. Accelerate to an ellipse that is a tangent to both orbits (and time it right)

Doing a Hohmann transfer involves first accelerating while in Earth orbit, at just the right time. This will put you into the elliptical orbit shown by the dashed line. You time it right with the orbits so that when you get to that outer point on the ellipse Mars is there. That is what drives the launch window, timing Mars. Often the idea is then you accelerate to circularize your orbit and match Mars' orbit. However, Mars has an atmosphere. It is pretty thin but adequate for entering the atmosphere either to be captured into an orbit or to descend to the surface. (Mars probes have done both.)

That means we only need a little bit of fuel to land, not to catch up. Starship will use the atmosphere to go into orbit. It will then select the right time to head for its landing site and fire rockets just a little to slow down. This will make it enter the atmosphere where it will glide in until the last minute where it will flip vertical and fire rockets just at the last minute to set it down gently on its landing pads.

In the case of Earth to Mars, the transfer between planets with this minimum energy (minimum fuel) approach is about 259 days and requires a delta-V from orbit of about 3.9 km/sec. Since we are using aero-capture, we don't need the second boost to match Mars' velocity.

However, for slightly more propellant usage, we can reduce this to 180 days as pointed out by Robert Zubrin and get the bonus of a free return if needed.

The delta-V for the 180-day transfer is about 4.3 km/sec. It just means you time it right, leave a little later, and intercept Mars a little earlier in its orbit. It also means the crew is exposed to 79 days less radiation during the trip. While the total risk for the entire trip is small, at about a 1% increased chance of cancer in the next 30 years (from about 20% to 21%), it is almost certainly a worthwhile trade-off to cut days off and reduce the risk. We reduce radiation and get 79 days more on Mars to do something productive. The same is true of the return journey.

Now to explain a free return for anyone who is not familiar with it. When Apollo 13 had terrible problems, they used a free return trajectory to get back. It is a path where if you don't deliberately alter your trajectory to be captured by the moon in their case, or Mars in ours, you will eventually come back to Earth. Apollo 13 did a slingshot around the moon. In the case of the Mars trip, the return flight would go past Mars and come back and take 18 additional months, but it gives you an option if things go terribly wrong. If everything is going well, you alter your path a little and get captured in Mars' atmosphere. If you do not alter your path that little bit, you will come back to Earth. That is a smart plan because we never know what can go wrong.

I should add that I did not confirm the math on the free-return trajectory. It is true that an orbit with a total transit time of 2 years, going out past Mars, should bring you back to Earth when it is right where you left it. However, the close pass by Mars would alter this path, so I am unconvinced that you are still on this trajectory. Either way I like the idea of shortening the transit time to stay on Mars. Let's assume 180 days for now and save the free return calculations for SpaceX. It is very likely there is a free-return option but it might involve a tight trajectory around Mars to slingshot back this way. The details of precise calculations like that are quite complicated. Real models include not only the sun and Mars but other planets in the calculations because they alter your path just a little. Even that model ignores the effect of the rest of the galaxy but is close enough with perhaps a mid-course adjustment.

As a college professor of mine once said, "Mother Nature applies all of her laws all of the time. The only reason we can solve anything is because we can ignore many of the forces as trivial most of the time." That is comforting to an engineer but probably sends chills up the spine of some scientists. I am the engineer. If I can calculate real-world solutions within a percent or two, I can build products that work.

In both of the cases of the Hohmann transfer and our planned 180-day transfer, I said 'about' on the delta-V. This is because Mars has an elliptical orbit. It is sometimes closer to the Sun and us than at other times. This means the exact dynamics of the transfer change from one launch window to the next. This changes the delta-V required, but not by a lot so let's use these numbers for now. It actually varies from about 3.6 to 4.6 according to Godard posted trajectories, but this is in the noise if it just means we add a little more propellant from tankers.

Therefore let's assume the transfer is 4.3 km/sec plus about 0.655km/sec for the mostly aero-entry. Let's add in 10 tons of margin propellant.

That puts us at a mass ratio of 3.774 plus our 10 ton margin. Using a mass of 120 tons for the ship and 120 tons of cargo, that means we need 676 tons of propellant less whatever residual is left from launch. That is just over half the tank capacity, but it is all we need and we probably don't want to land with a lot of extra mass, which would make the landing more difficult or even impossible.

We take on tankers as Elon Musk described in his update in 2021. The tankers fly up to orbit, mate tail-to-tail with the Starship, and transfer fuel. This amount of propellant is probably about 5 or 6 tankers based on payload capability.

Once the ship is refueled, we are headed to Mars. Our computer just has to fire the engines at the right moment in our orbit to increase our speed around the Sun.

Staying on Mars

The way the launch windows line up, doing 180 day transfers both ways, our stay on Mars will be about 550 days or about a year and a half. Unlike the moon trips, this is a long stay offering the chance for plenty of exploration and science. It also provides plenty of time to refill our tanks using the fuel factory that will be described in chapter 6.

What will the crew do during this time? They will spend some of their time tending the fuel factory and adjusting solar panels. They will also have to collect regolith using heavy equipment like perhaps a front end loader and a backhoe that operate on Tesla batteries. They will need to collect about a ton and a half per day. But the crew can do more than just this if there are several crew members.

The first human mission will probably live in the ship. It has excellent recycling capability and does not require any construction. It is possible to take along some form of habitat and later missions certainly will, especially when they plan not to return to Earth. The first might or might not.

Early missions will explore for water. A better source saves power and volume. The first missions will probably boil regolith (Martian soil) to get the water out of it. Some locations on Mars are very rich in water ice embedded in the regolith, on the order of 50–60% water ice by weight or higher. However, the ultimate goal would be either sheet ice or better yet a geothermal well. It is believed Mars has warm or even hot water some places underground. A find like this could not only be a source for unlimited water but enormous amounts of power too. There will definitely be testing done looking for a rich water source. Common methods currently are ground penetrating radar and resistivity surveys. Devices are sold online using resistivity surveys that claim to be able to detect up to 500 meters deep, but they involve electrodes that must be inserted into the dirt.

Some of the ground penetrating radar systems look almost like a lawn mower with the radar system mounted between the wheels and a computer screen display for the operator pushing it, making it mobile. Why not integrate that system into the rover, where the radar system is on the belly of the rover (or a small trailer) and the crew has a display of the output?

I cannot verify the accuracy of either but this is an opportunity for SpaceX to test the product claims or bring in an expert to optimize the system. Something like a mobile system on the rover would allow the crew to cover huge areas over an eighteen month stay.

Many later missions might also look for other resources. Mars was once volcanic. Volcanoes can melt down metals and create rich deposits. I don't mean they will randomly dig for gold or other metals, but scientific instruments might help locate these deposits over time. Many metals could be useful, whether for building structures or the beginnings of their own industrial production eventually. They will also want to study Mars and understand more about it. Science has to be a part of an expedition so grand. However, basic survival and finding resources is certainly a first priority.

During a year and a half stay, they will have time for significant exploration. A rover or two with Tesla batteries could provide them very broad mobility and be charged daily from all of the solar panels they are going to bring.

Staying Supplied

Consumables are an important part of our payload. We have to take along food, and perhaps oxygen and water. We can do some recycling, but how much? That is the question. For an answer let's look a little at the Space Station for something close to state-of the-art today. They have a pretty elaborate system of recycling and purifying things that are recovered.

First, let's look at water. The NASA standard on water consumption is as shown in Table 2[3]. Note that where I use kg/crew/day this means each crew member consumes this amount each day of the mission. Also, these rates are based on years of consumption rates aboard the International Space Station (ISS), so they are quite accurate on average.

Water Required	
Usage	**Kg/crew/day**
Drinking	2
Food Rehydrate	0.5
Medical	0.05
Hygiene	0.4
H2O Flush	3.2

Table 2: Crew water usage

This is quite a bit. However, nearly all of this can be recovered by an ISS level environmental control and life support system. Table 3 below shows the level of recovery[3]. Note that the spacecraft is a completely enclosed environment. Nothing escapes other than some tiny fractional leakage through a hatch seal or waste that we cannot recover and probably vent to space.

Water Recovery			
Source	**Kg/crew/day**	**Recovered**	**Loss/crew/day**
Crew Latent	1.87	100%	0
Urine	1.49	85%	0.2235
Flush	0.25	85%	0.0375
Medical	0.05	100%	0
Hygiene	0.4	100%	0
Total:			0.261

Table 3: Water recovery by source

Based on this level of recovery, our requirement went down from 3.2 kg per day to 0.261. That says we might be a little short. However, we actually will have a slight excess as we will see from our food.

So, let's talk about food. The NASA standard for food is 1.831 kg per crew per day [3]. This food is freeze-dried. However, it still contains 28% water by mass. This gives us 0.513 kg per crew member per day of water added to the system which yields a net surplus of 0.25 kg of water per crew member per day, so we have to take a lot of food, but we only have to take enough water to start the system up and to have extra for contingencies. If we assume 30 days' worth to start the system up and for contingency, that would be 96 kg/crew member.

What about oxygen? According to NASA, a person will consume 0.82 kg per day. In the process, they will generate 1.155 kg of CO2. What do we do about that CO2? What the ISS does is they have a small Sabatier reactor, which we will discuss in detail in chapter 6 because it is how we will produce fuel. However, in this case, we just want to scrub the CO2 out of the air and recover the oxygen. The Sabatier reactor converts CO2 and hydrogen into methane and water. The methane in the case of the ISS is vented overboard, although we could capture it if we choose to. It is produced at a rate of 0.42 kg per crew member per day. That is a pretty small number compared to what we will generate on Mars, but it is almost half a ton on the way there for a crew of 6.

You might ask, where do we get the hydrogen? The answer is electrolysis, just like the ISS. We use electric current to split the water produced into hydrogen and oxygen. The oxygen is used to breathe and the hydrogen is fed back into the Sabatier reactor. The only issue is half of the hydrogen goes into the methane, not into the water. However, the 0.25 kg/day per crew member excess water we have is almost exactly a balance to make up for the lost hydrogen if we feed that into the electrolysis unit too.

Sounds like a perfect system. We get everything back, right? Well not quite. In the Sabatier reactor, about 2–5% of the CO2 doesn't react. We can condense out the water and recover that but let's assume we are not going to the point of cryogenic chillers to separate out the CO2 and methane. That means we need a little bit of water to make up for the losses. It is almost trivial but it is not zero. It is about 5% of the water produced and half the corresponding amount of hydrogen that is vented rather than consumed. Half, because remember that we were already assuming half the hydrogen is used up as methane so the net loss is only 2.5% of the total. This corresponds to 0.05 kg of water per crew member per day to make up for the 5% loss in the incomplete reaction. The end result is shown in Table 4.

Net Requirements After Revovery			
Item	**Required(kg/crew) 910 days**	**Contingency(kg/crew)**	**Remarks**
Oxygen	0	150	Oxygen recovery is 95%
Food	1666		NASA standard
Water	46	96	Replaces lost hydrogen and oxygen
Hygiene Cons	72		
Wipes, bags, etc	182		
Personal, Misc	79		
Human	75		165 lb average with 50/50 gender mix
Total	2120	246	Total is 2.37 tonnes per crew
Note: Contingency is 30 days water, 180 days 02 for system failure			

Table 4: Net requirements for mission with a closed loop recovery system

Note that this includes the person and their personal effects.

We have a slight loss of O2 in the form of CO2 not reacted. However, we also had excess water we are using for electrolysis. The net result is about 0.181 kg per day excess oxygen or 132 kg per crew member excess over the 180 days to Mars plus the 550-day stay. We would want to capture this for future use on Mars.

The numbers in table 4 are what we will use per crew member. For contingency, I used 30 days water. This is a closed system. We should have multiple condensers (de-humidifiers) so it is not credible that we can't recover most of our water even if something else fails. Even though our system should be very reliable, I couldn't see leaving on a 180 day flight without a backup on oxygen. One-hundred-fifty kilograms isn't much to pay to be fail-safe. We even have oxygen tanks we are taking for fuel production. That could be where we store this oxygen for the trip if we don't want to bring additional tanks.

Note that once on Mars we should have all of the water and therefore oxygen also that we need once we start our fuel production operations. That is a reason for not taking more oxygen than just the trip out as a backup.

Just thinking of contingency planning, what if we got to Mars and were having trouble producing our return fuel fast enough to get back in the next window. This could be due to having a much worse water source than we expected, for example. We would have to wait over two years for the next window. However, another ship will be arriving about 49 days after we were supposed to be back on Earth. We would be in communication. We could easily relay our situation and the next ship could bring more food or whatever is needed. We would only have to last 49 days beyond our planned mission. We don't really have to bring years of supplies, although that is a decision the real mission planners will make. When we get to our mission plan we actually assume we want to take enough to last until the next window but that is pretty conservative.

Getting Back

One of the main things we are taking with us on the first trip is the fuel factory to get us back. This factory is almost half our payload by mass. However, it can use Martian water and CO2 to make all of the propellant we need to get back, which is almost 15 times the mass of the factory. This factory is described in detail in chapter 6 and does not require any new technology. In fact, other than solar cells it is based on technology that is over 100 years old.

In order to return from the surface of Mars, we need a delta-V of about 4 km/sec to get to orbit, another 3.1 to get back in a similar transfer orbit to the one we arrived on, and another 0.1 or so to land. In order to do another near-Hohmann transfer of 180 days when we are in orbit around Mars, we fire the engines when we are on the Sun side of the planet. This accelerates us out of

Mars orbit but actually slows down our speed relative to the sun. This causes us to fall back toward the Earth on the ellipse shown previously (moving closer to the Sun).

If we do it like the previous transfer, we will use just a little extra propellant as compared to the Hohmann transfer to get us back to Earth in 180 days. That number is the 3.1 km/sec already quoted. Once we get back to Earth we do an aero-capture in Earth's atmosphere, entering very much like the Space Shuttle did until we get close to the ground. The rocket then flips on end to turn vertical and be caught by the chopsticks. On the way in, we probably do the loop around like the Shuttle did, which allows us to adjust for any inaccuracy in our entry path, as in do a bigger loop or smaller loop as needed to adjust.

The payload we are taking to Mars, including solar panels, the fuel factory, rovers, and extra supplies, all things that will be useful for the next ship, will be left on Mars for the use of future missions. This is precious cargo for Mars, never to be returned to Earth. The only thing going back is the crew, provisions for them, and perhaps some Mars samples for analysis on Earth.

The way the windows line up, the next ship will arrive about 49 days after we get back to Earth, give or take a few days, depending on whether both of us launch a few days early or late in the launch window. That means one wave of ships would never see the next one because they will be in transit at the same time, the previous one going home while the next one is on its way.

Our return cargo will be on the order of 4 tons. This will only consist of the crew, provisions for them, and any Mars samples, as described previously. The delta-V of 7.2 puts us at a mass ratio of 6.82. Using 124 tons as our final mass (ship plus cargo), that means we need 722 tons of propellant plus perhaps a 10 ton margin again for a total of 732 tons. That is a lot of fuel but still less than two-thirds the capacity of Starship.

I would like to offer an analogy on this approach versus the traditional approach. My in-laws lived in Virginia and we lived in central Florida. It was about 750 miles (1,210 km). The traditional or SEI approach to travel was to take a semi-truck and drive it until it ran out of gas. They abandon it there. In the back it has a pick-up truck and a car. They used the pick-up truck to pull the car until the pickup ran out of gas and abandoned it. Then they drove the car and arrived at grandma's house. With this approach, we would never go see the grandparents because it costs $300,000 one way!

The approach just described in this chapter is “Let’s just drive the car and stop for gas a couple of times.” With the ‘new’ approach, it costs maybe $300 in gas and vehicle costs to go to Virginia and it happened a couple of times a year.

We would never imagine throwing away vehicles for travel on the Earth after one use and couldn’t afford it if we did. Why did we ever in space outside of perhaps early testing?

The Calendar

For a Hohmann Transfer, or in our case just a slightly shortened 180-day version, as described already, the planets have to be in alignment, such that Mars will be where you will in 180 days or Earth will coming back. Since Mars is farther out, it takes about 687 days to circle the sun. The Earth takes about 365.25 days to do a similar orbit closer in. The math says they line up just right every 2 years and 49 days. That means a launch window opens every 2 years and 49 days. You don’t have to hit it exactly but it has to be pretty close or you need more fuel. The difference is you just aim a tiny angle differently, or do a mid-course correction, if you leave a few days early or late. However, months late or early doesn’t work at all, without massive fuel usage to catch up and unmanageable approach velocities when you get there.

For the optimal return, the stay on Mars is about 550 Earth days. That means a round trip is 2–1/2 years. It also means the return trip will be in transit during the next launch window, so they will pass each other, well, on opposite sides of the Sun, but in transit at the same time. This however, means they are never on Mars at the same time needing to share resources. The refueling factory left by the first mission can be used by the second mission. The calendar is depicted in Figure 8. I arbitrarily chose the window at the end of 2026 to start the missions. That is a guess but perhaps close, but the same principle applies even if we start at a different launch window, such as starting where mission 2 does, instead of mission 1.

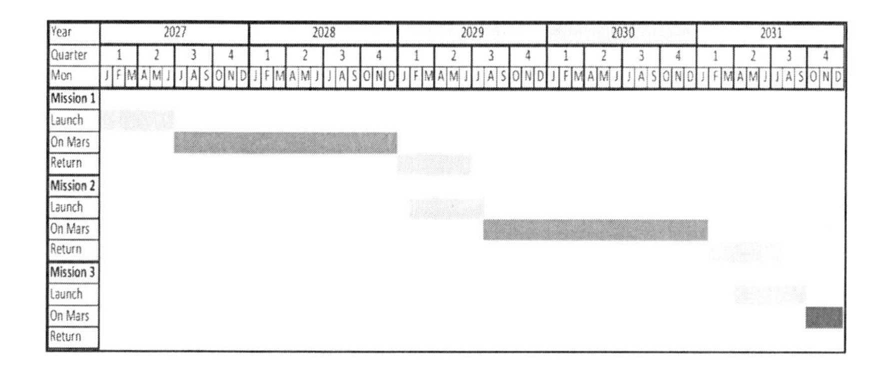

Figure 8: Mars mission timeline

The round trip is 2.5 years but the launch window is every 2 years 49 days. This means sequential missions are not on Mars at the same time. (The dark bars are time spent on Mars. The light bars are transit time).

The frequency of launch window also slows any efforts like colonization if you can only go about every 2 years. There should be some level of demonstration before we send large numbers of colonists, so it may not be as fast as Elon Musk says (a million by 2050) but I love his vision. I think he is trying to shock people or at least wake them up with numbers like that.

As we will see later, a thousand or a couple of thousand colonists by 2050 or so is definitely possible if we go full speed on colonization, but even that level takes a LOT of infrastructure. Perhaps they could accelerate it faster but when we get to city levels we need a power source beyond just acres and acres of solar cells that work half as well as they do on Earth and don't work at all at night. At the city level we also may need other resources, like underground malls where people can trade for things they want or need. We probably need utilities and a power grid. We also need lots and lots of greenhouse acreage. One solution on power is if we found a geothermal source we could produce megawatts of power from the contained energy and it runs 24/7.

(The correct term is actually areothermal. Geo is the prefix for Earth. Areo is the prefix for Mars. However, everyone understands geothermal so I will use it.)

The Environment of Mars

For me, it is far better to grasp the Universe as it really is than to persist in delusion, however satisfying and reassuring.

Carl Sagan, scientist and author

I want to die on Mars, just not on impact.

Elon Musk, entrepreneur and visionary

Planetary Characteristics

Mars is 1.52 times as far away from the Sun as the Earth. This means it is colder and the year is longer. A Mars year is 687 Earth days. A Mars day however is almost the same as ours at 24 hours and 40 minutes. If you have seen any other source that says 24 hours and 37 minutes, the difference is a rotation is 24 hours and 37 minutes. However, the planet is moving around the sun. A day relative to the sun is slightly longer at 24 hours and 40 minutes.

The Earth tilts 23 degrees on its axis, giving us our summer / winter cycle. Mars is very close at 25 degrees. This means it's tropical and arctic regions extend 25 degrees from the equator or pole respectively, slightly farther than Earth's. The term tropical implies warm to most people. On Mars it is warmer there, but far colder than the kind of temperature averages we see on Earth in tropical regions. Just like Earth, the summer / winter daylight hour swing is greater as you go toward the poles. In the arctic region half the year is in constant night and half the year is in constant day, though with a relatively low sun angle, just like on Earth.

Figure 9 shows the Mars rover Curiosity on the surface of Mars. This shows what a lot of the surface is like—rocky and somewhat barren.

Figure 9: NASA's Curiosity Rover on Mars
Credit: NASA

Because Mars is much smaller and lighter, its gravity is only 38% of Earth's gravity. If you weigh 150 pounds (68 kg) on Earth, you will only weigh 57 pounds (feels like 26 kg) on Mars. (I used English units only because technically a kilogram is a mass, not a weight.)

We know that zero gravity has some detrimental effects on humans during prolonged exposure. These affects include gradual bone density loss and muscle atrophy. Whether 38% gravity is enough to maintain bone density and muscle strength remains to be seen but it certainly has to be better than zero g. Perhaps it won't even matter if you colonize Mars. You will always be working under 38% gravity and therefore require less strength and bone density to perform tasks and avoid injury.

Atmosphere

Mars has far less atmosphere than the Earth. The atmospheric pressure on Earth is 14.7 psia at sea level or 1013 mbar. Mars' atmosphere actually varies some but it is between about 6 and 8 mbar most of the time with seasonal excursions just outside of this range. Those numbers also vary with elevation, just like on Earth. Since Mars does not have any surface water, I will refer to 'sea level' values as the mean elevation. The atmosphere is actually about twice this pressure in the bottom of Valles Marineris, Mars' deepest canyon, which is over 4 times as deep as the Grand Canyon at 7 km (23,000 feet) deep.

While our atmosphere is 78% nitrogen, 21% oxygen, and about 1% other gases, Mars atmosphere is 95.32% CO2, 2.7% nitrogen, 1.6% argon, and less than one percent remaining that includes oxygen, carbon monoxide, water vapor, and a couple of other trace gases like methane and neon. To say is it not breathable is an understatement. Our blood would boil at the near vacuum pressure and we certainly can't tolerate nearly pure carbon dioxide. It is a very inhospitable environment without a space suit. Does that mean we can't go? Absolutely not. Does it mean there will be challenges? Yes it does.

The good news is that it has resources we can use. Carbon dioxide (CO2) and water are available and can be turned into rocket fuel and oxygen as needed. One of the Mars rovers produced a small sample of breathable oxygen from Mars' atmosphere. We just have to apply a little bit of engineering to produce what we need.

Mars Atm:	
CO2	95.32%
N2	2.70%
Argon	1.60%
O2	0.13%
CO	0.07%
Water	0.03%
Ne, Kr, Xe, O3	Trace PPMs

Table 5: Mars atmosphere

Temperature

Something interesting is that we can calculate the predicted temperature of a planet. Any observed difference is absorption by the atmosphere, also known as greenhouse effect. We know the Sun's radiation at a given distance. For Earth it is 1367 kW/m^2 less 30% reflection (clouds, etc.) for a net of 957 W/m^2. For Mars it is 592 W/m^2 less 16% reflection for a net of 497 W/m^2. This heat input must be balanced by the outward radiation (mostly infrared) by the planet based on its temperature. The radiation of any body is determined by a formula based on temperature to the fourth power (a few more degrees radiates a lot more heat). The heat from the Sun radiates on the face of the planet. The planet radiates on all sides of the planet. The area of the face is πr^2. The area of the entire surface is $4\pi r^2$. This simplifies it in that the area of output radiation is 4 times the area of input radiation. This provides the formula for the radiation balance that allows us to calculate the predicted temperature.

$$\textbf{Temperature} = (\textbf{Net Solar Flux}/(4*\textbf{Stefan-Boltzmann Constant}))^{0.25}$$

Temperature is in degrees Kelvin. Net solar flux is in W/m2 as described in the previous paragraph. The Stefan-Boltzmann constant is 5.67×10^{-8} W/m^2K^4. If you don't like the math, just bear with me for the results.

The predicted temperature and actual observed is shown in Table 6 for Venus, Earth and Mars. One interesting thing this shows is that Mars has almost 17 times the CO2 that Earth has but only a tiny fraction of the greenhouse effect. This is because most of Earth's greenhouse effect is water vapor and ozone.

I am not saying we haven't warmed the planet a little bit, just that anything we have done is a small addition to a significant natural and useful greenhouse effect. Without the natural greenhouse effect, our oceans would freeze over. Venus on the other hand is an extreme of the greenhouse effect. It has about 92 times the density of our atmosphere and the atmosphere is CO2 and Sulphur dioxide. Venus is a raging inferno.

	Venus	Earth	Mars
Distance(AU)	0.72	1.00	1.52
Solar Radiation(W/m2)	2637	1367	592
Reflection	75%	30%	16%
Net(W/m2)	659	957	497
Radius(km)	6052	6278	3396
Calc Temprature(°K)	232	255	216
Actual Average Temp	735	288	218
Greenhouse Delta	503	33	2
Density CO2	93.288	0.00042	0.007
Radiation Rate	66190	1560	512
Absorption	99.00%	38.67%	2.97%

Table 6: Comparative solar radiation and properties of planets

For reference, 273 K is the melting point of water ice.

The real takeaway for our purposes is that Mars is very cold. The planetary average is 218 Kelvin which is -55ºC (-67º F) and about 70ºC (126º F) colder than Earth. This does vary with latitude. A calculation at the equator yields an average temperature of about 232º Kelvin which is -41ºC (-42º F), though daytime temperatures as high as 21º C (70º F) have been recorded at the equator of Mars. With such a thin atmosphere, the day to night variation can be extreme, on the order of 50–100ºC (90–180ºF) difference. These temperatures are important when it comes to heating living quarters or a greenhouse or getting rid of excess heat in fuel production. Note that this variation is in the thin atmosphere and perhaps right on the surface of the regolith (soil), not deep in the ground. The actual ground, especially down a meter or so, does not have these kinds of swings overnight, except in arctic regions.

The average temperature varies by latitude and can be calculated. However, the day to night variation becomes truly extreme within Mars' Arctic Circle. Just like here on Earth, areas within the Arctic Circle have half a year of round-the-clock daylight and half a year of round-the-clock darkness.

Radiation and UV

The Earth has an atmosphere with an Ozone layer. It filters out most UV rays. Earth also has a thick atmosphere that significantly reduces radiation that comes from cosmic radiation and from the Sun. Mars has neither.

Let's first talk about UV rays. The level of UV rays on Mars is fatal to crops. They also are quite bad for humans, ranging from sunburn to skin cancer. Crops cannot be grown out in the open on Mars, like in a greenhouse UNLESS we can filter out the UV rays, hopefully without filtering out the visible light that plants need. There is a solution to that which we will discuss in Chapter 8 when we get to infrastructure requirements. It involves the filtering materials used for the greenhouse.

The next topic is radiation. First of all, let's start by saying gamma rays are the issue. You can reduce them but it is impossible to take it to zero. On Earth, thanks to our thick atmosphere, the average dose is about 0.24 rem/yr. On Mars, it is about 40–80 times that. The variation is primarily elevation, which determines how much atmosphere is over your head. Lower elevations are better. A lower elevation is also good for atmospheric thickness and for warmth, both of which are beneficial for us. We would choose a low elevation anyway, but radiation adds another reason to look to colonize at lower elevations. Figure 10 shows the radiation levels on Mars by location. The original was in color but it is printed here in black and white. That makes it difficult to distinguish the extremes from each other, but we would want a medium grey area in the northern hemisphere. They are toward the lower end of the scale. However, other considerations may drive, which area? For example, we don't want to be in a canyon or crater we can't drive a vehicle out of.

Note that the southern hemisphere has generally higher elevations and therefore higher radiation.

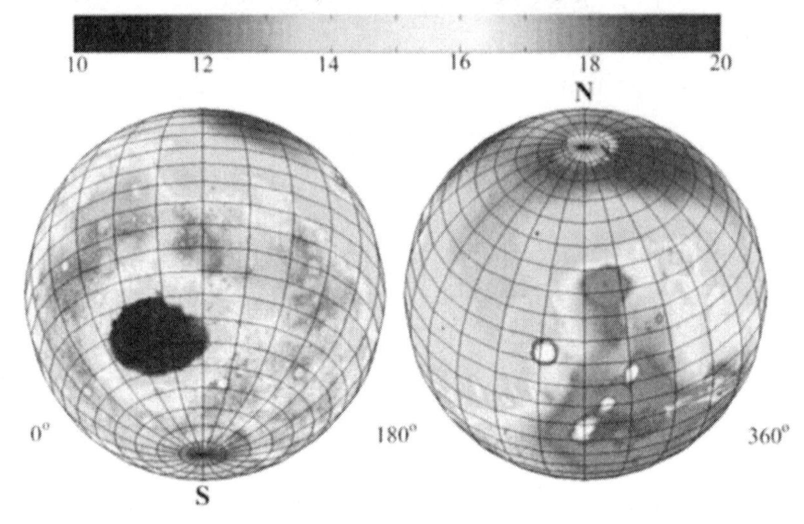

Figure 10: Cosmic radiation levels on Mars
Credit: JPL

Later, we will discuss what a safe level of radiation is and what humans can tolerate. For now, suffice it to say that levels on Mars are quite a bit higher than on Earth, but they vary by region, mostly by elevation.

Dust Storms

Mars' orbit is elliptical. Therefore, it is warmer at one point in its orbit than at the opposite side of its orbit, because it is closer to the Sun. This corresponds with summer in the southern hemisphere. As a result, it gets warm enough that huge volumes of CO2 that have frozen and been deposited in the southern polar region boil off from the Sun's heat. This results in a +/- 20% swing in Mars' atmospheric pressure from its mean of about 6.5 or 7 mbar at mean elevation ('sea level'). Mars has dust storms most years. This boiling-off of CO2 in the southern hemisphere coincides with (causes?) the annual dust storms, though surface heating also helps to drive them. Note that during winter, the cycle reverses itself and the CO2 refreezes over the poles at about -123 C. (Yes it gets really cold at the poles). Hence the pressure swings.

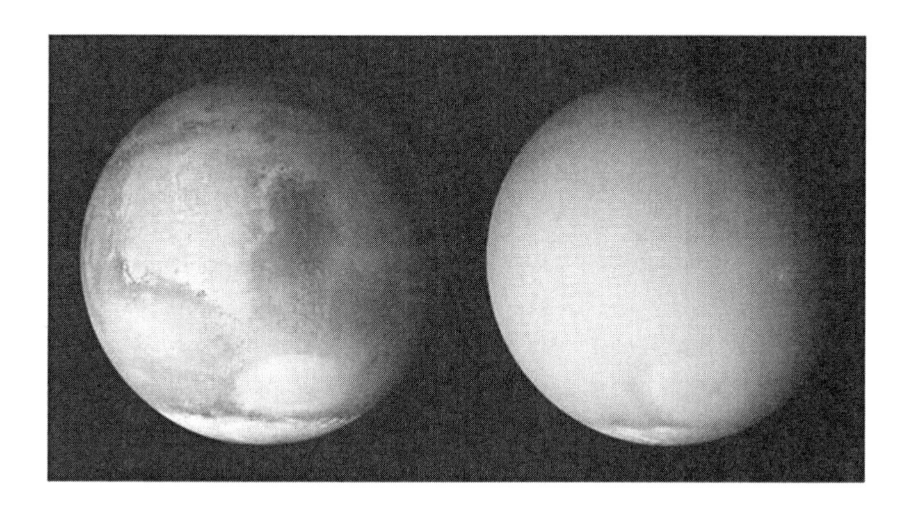

Figure 11: Mars normal appearance and appearance during a global dust storm
Credit: NASA

These storms can be continent size but every third Mars year (5.5 Earth years), they usually turn into a global event[4], as shown in Figure 11. It is not fully understood why, but it is believed this may be that it takes that long for the global cycle to completely reverse itself back to the conditions that lead to global coverage.

These storms can make Mars look like a smooth globe over part or all of it because the dust blocks the view of the surface. From Mars' surface, that means it blocks most of the sunlight. Blocking the sunlight is an issue for solar power, as we will discuss later.

The dust from the storms not only temporarily blocks the sun; it also winds up on everything. However, with humans present, this is only an added task, cleaning the panels, not an irreversible event.

Note that winds in the storms are not significant on the surface. Even though the wind speed is on the order of 100 kph (62 mph), the atmospheric density is so low that this would feel like a gentle breeze of 8 kph (5 mph) on Earth. This would not strand a vehicle on Mars or stop any other human activity. However, the dust is very fine and tends to get into everything so a human may choose to stay inside rather than later having to get all of the grit off the suit and out of any joints in it.

For anyone that saw the movie *The Martian*—while a pretty good movie and an interesting story—it is not based on facts, as far as the storms. The storms on Mars could not blow people down or launch an antenna through his suit. The air is far too thin. It also would never tip a rocket over. However, there are parts that make a lot of sense. He was frequently cleaning dust off his solar panels. That could be needed, especially after a storm. He grew potatoes in his own feces. That could work. There is no organic material in the soil and little or no nitrogen. We will have to add something to the soil. We may bring fertilizer, but we will probably add our own waste as organic material to enrich the soil too. Urine actually is an excellent fertilizer.

I use the term soil here, but technically the term is regolith for non-Earth dirt that lacks the organic materials. Soil is the kind of rich dirt we use on Earth to farm. However, regolith may sound odd to you at first but is the correct term for Martian dirt.

Storms have affected landings. Russia lost two unmanned spacecraft due to coming directly in during a storm, as opposed to orbit first, then land when you choose to like an American probe that waited out the storm in orbit. However, the Russian craft were parachuting, so they were dragged with the wind rather than landing gracefully. An automated Starship with radar ground level and terrain sensing might not have this issue. Remember, it is the equivalent of an 8 kph (5 mph) crosswind, which would be considered very light in aviation. However, the crew might choose to wait in Orbit until they can see the ground. There may be very little solar power available down on the surface during the storm, but there is in orbit.

If a Starship is caught in a dust storm while on the ground, they will probably have very little solar power. I say little because while the globe looks mostly a uniform reddish in Figure 11, you can actually see dark and light spots through the dust similar to the dark and light spots in the image on the left. This says solar energy might be reduced to perhaps five to ten percent but not zero. Even a little would be helpful; otherwise, anyone on the surface is on batteries until the storm dies down, which could be weeks. They probably would shut down the fuel production described in the next chapter until they have significant solar power again. However, if it was even fifteen or twenty percent, they would continue producing fuel, but at a reduced rate.

Water

Mars is believed to have once contained vast oceans. It still has a lot of water. Some of it is above the surface in polar regions but most of it is underground and extends at least as far south as 45 degrees latitude. An example is shown in Figure 12. This is the Korolev Crater. It is located at 73 degrees north latitude and 165 east longitude. There are others like it and polar ice caps. This one source is estimated to be 2,200 cubic kilometers (530 cubic miles) of ice. That is about the volume of Lake Ontario and Lake Erie combined. However, the ice is 2 kilometers down in the crater and within the Arctic Circle. The crater is 81.4 kilometers (51 miles) across, just to offer scale to the picture. The left side apparently faces south, since the right inside of the crater wall, which would see sun, has very little ice on it. This is also true of the upslope outside of the crater on the left.

Figure 12: Korolev Crater at 73 degrees latitude is one of many places with water ice in polar regions
Credit: ESA

The polar ice cap, as shown in Figure 13, is approximately 2 kilometers thick but is riddled with dust. This is believed to be caused by the dust storms initiated at the South Pole. When they occur, it is winter at the North Pole and some of the CO_2 released would condense here. However, this cap has been determined to contain far more water ice than CO_2 ice based on spectrographic analysis.

While there is extensive exposed water ice in polar regions, it cannot survive on the surface at lower latitudes. However, NASA surveys have demonstrated that there is water ice just below the surface at least as far south or north from the poles as 45 degrees latitude.

Figure 13: North polar ice cap captured by the Mars Global Surveyor
Credit: NASA

The History of Mars

Mars was once a warm, wet planet. Scientists believe it had liquid water in vast oceans a couple of billion years ago, probably mostly in the northern hemisphere. That would imply it was a lot warmer and had a much more substantial atmosphere.

It is believed that Mars lost its magnetosphere at some point. If that happened it would probably start very slowly shedding its atmosphere, which in turn would reduce its greenhouse effect. It probably froze during a period of perhaps millions of years as its atmosphere either froze on the surface or was lost into space. Probably because water has a relatively high freezing point, compared to gasses in the atmosphere, much of it froze. That would be why there are still huge reserves of water on Mars in the form of ice, often mixed with regolith (dirt) or some of it may be sheets of ice just covered by dirt.

Mars has gotten very cold since then and all of the remaining water not at the poles is contained as ice in the regolith and probably underground reservoirs and ice pockets. It also has hundreds of thousands of cubic kilometers of solid water ice (and some CO2 ice) at the poles. The portions of the atmosphere that didn't condense out and freeze were probably lost. Mars no longer has a strong enough natural magnetic shield to protect its atmosphere from the effects of the Sun like Earth does. Even now it is losing tiny bits of its atmosphere over time.

Figure 14: The South Polar ice cap is shown clearly here, as well as the 25⁰ tilt of Mars
Credit: NASA

It is believed that Mars was very geothermally active when it had liquid water on the surface. Scientists have evidence now that it is still geothermally active and contains warm underground liquid water[5]. The evidence is for water underneath the polar cap but if it is there, it probably exists elsewhere too. The pole is about the least likely place to have liquid water. They only know it is there because it is causing ripples and effects on the ice that wouldn't be possible without a layer of water underneath. We haven't searched the rest of the planet for underground rivers, but will someday.

Now we have looked at the environment on Mars. Let's look at how we can take advantage of the resources it has.

Using the Resources for the Production of Fuel

Creativity doesn't wait for that perfect moment. It fashions its own perfect moments out of ordinary ones.

Bruce Garr Brandt, artist

Creativity is what this section is all about. The people at NASA, back in 1989 while doing the ninety-day study, didn't think out of the box on how to NOT carry a huge, fully fueled rocket all the way to Mars so that the astronauts could get back. They assumed: carry everything with you. The result was a monstrous rocket and a monstrous infrastructure. In his book *The Case for Mars*, Robert Zubrin explained this concept of making fuel on Mars. Elon Musk has latched onto it. In this chapter, I will flesh this concept out on a larger scale for a ship the size of Starship.

The Overall System

This system as shown in Figure 15 will produce 732 tons of liquid methane and oxygen in the correct mixture ratio in about 480 Earth days. This is the calculated amount required for a return trip back to Earth from the surface of Mars. This allows a 70 day margin over the standard stay of 550 days to allow for start-up time and possible time lost due to dust storms, plus some margin in case we are off a couple of percent in one of our estimates. If we are off 2% in power production for example, we finish 10 days late and are still ready in plenty of time for launch.

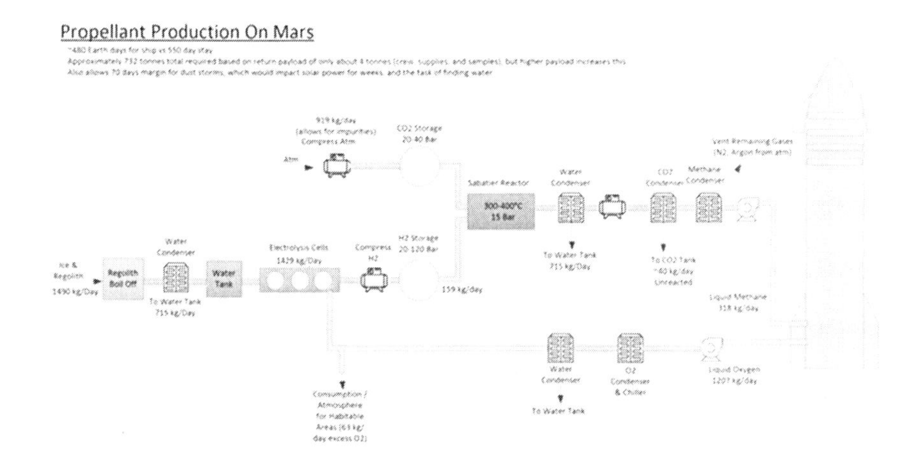

Figure 15: Overall fuel production system

It should be noted that the proposed architecture simply compresses the atmosphere and feeds it into the Sabatier reactor, along with the hydrogen. This is not what some others are proposing. They want to condense out the water, then the CO2 and feed only pure CO2. That is not necessary and uses a lot of extra power. It has been demonstrated that biomass waste can be used directly to feed a Sabatier reactor[6]. It is nowhere near pure CO2. The inert gases had no effect on the process other than having to readjust quantity for how much CO2 and hydrogen are actually going in. There is no need to remove the Argon, Nitrogen, water, or O2 (all single-digit percent or lower levels). The water is a natural product and therefore captured. The O2 will be consumed in the reaction. The Argon and Nitrogen are inert and have boiling points below methane and can simply be vented to the atmosphere after condensing out what we want, or they could be captured and used as we will later discuss.

Power

The first consideration is how we create power, and a lot of it. There are at least 2 options. One is solar; the other is nuclear. Many NASA programs have chosen nuclear, especially as we go further from the Sun. While the nuclear option is more efficient, especially for power intensive missions, since it is compact, light-weight, and will produce power for years regardless of local conditions, nighttime, or storms, that is not what is going to happen. First, SpaceX is not going to wait years for a nuclear development program even if they can get plutonium or enriched uranium (and perhaps can't or can't without years of government red tape). Secondly, Elon Musk has stated that they are going solar. So let's look at solar on Mars.

In order to produce fuel on Mars, we need power, and actually quite a bit of it. The capability to do this is a lot less mass than the propellant generated, but it is not trivial. It is almost half of the payload capacity.

Solar cells will work on Mars but not as well. Mars has about 52% of the energy available per area compared to Earth at 497 W/m^2 versus 957 W/m^2 so the power level is reduced but they will work. The reduced power is because it is 1.52 times as far from the Sun as the Earth is, and power goes down as the square of the distance. It should be about 2.3 times less based on distance but Mars doesn't have much of an atmosphere to reflect it, so the ratio isn't quite that high. The lower power just means we have to take more panels.

It should be noted that Mars is in an elliptical orbit, meaning it is sometimes closer or farther from the sun at different times of the year. The farthest point is near June 1st Mars equivalent. This means solar flux is not constant. It varies some, but how much is partly offset by the winter/summer length of day cycle in the northern hemisphere. How much, depends on latitude. For now we will use the average and address that issue when we start talking actual mission planning and what latitude we are going to.

One other factor is the angle of incidence. In other words what latitude is the panel at and how direct is the sun coming at the panel rather than going through a lot of atmosphere. Mars doesn't have a lot of atmosphere, so this seems that it would be negligible. However, it has dust in the atmosphere and that dust both absorbs some of the light and also diffuses light. The diffusion (scattering) can change frequencies and therefore cause some solar cells to be less effective and some to be more effective. This is shown in Table 2 where the NASA rovers Spirit and Opportunity measured the effectiveness of different solar cell materials based on sun angle (more atmosphere to go through).

The finding is that some cells are even more effective than a cell on Earth, relative to available power and some less. The range was about 81% to almost 104% of effectiveness compared to Earth, based on use of available sunlight[7]. The average ranged between about 95 and 98 percent, or a 2 to 5 percent loss as shown in Table 7. This data is for a relatively poor visibility period, as in dusty but not during a dust storm, so this should be about the worst we see other than actual dust storms. Dust levels tend to be quite low in the spring and summer. However, this is often followed by the dust storms in early winter.

Solar Cells on Mars:

	Sun Angle					
Cell Type	**33°**	**44°**	**55°**	**63°**	**76°**	**78.5°**
InGaP(top)	86.2	84.9	85.3	81.4	90.7	90.9
GaAs	94.1	93.2	93.8	93.1	97.2	97.2
GaAs(mid)	102.8	102.2	103.1	101.6	103.8	103.7
Si	100.4	100.5	100.4	100.5	100.4	100.5
Avg	95.875	95.2	95.65	94.15	98.025	98.075

Table7: Solar cell effectiveness by solar angle from Spirit and Opportunity rovers (90 degrees equals directly overhead)

Obviously, there is some measurement error in the data, as there is in all data. However, it is interesting that the last two tend to be almost flat and the first two appear to go down and then back up a little. Does performance actually improve for these compounds at more extreme Sun angles? That is not certain, but it definitely does not drop off radically at low angles based on this data.

Regardless, with this data and other in hand, any solar panel going to Mars will clearly be optimized for the Martian environment. It is unclear whether we will have very slight degradation, or perhaps even slightly improved performance. The answer may be slightly degraded. Solar panels use multiple layers and multiple materials to capture different wavelengths of the solar energy. However, two or three percent of the power produced is probably in the noise when we are making an assumption as to the efficiency of the panel.

For our purposes, let's assume a net efficiency including this minor loss of 25%. Commercially available solar panels range from about 15% up to or very near 25% efficiency. The 22–25% efficient solar panels are more expensive, but that is not an issue for our purposes. Our solar panels will almost certainly be custom panels tuned to Mars, so let's assume at least as good as the best panels you can buy for your home.

While cells have demonstrated 39.5% efficiency in a lab[8], these are fairly exotic cells with many layers and may or may not be scalable for our purpose. They could be heavy and perhaps difficult to manufacture on a large scale. I prefer not to assume world record performance, only the top end of the commercial market. However, this gives credence to the 25% assumption above being very achievable.

Based on these numbers, the maximum output from solar cells is 124 W/m2, about half of similar solar cells on Earth. This is with a direct ninety-degree solar angle. On Earth, we have solar panels with drives that keep them pointed directly at the sun (at solar farms, not on rooftops). These drives are heavy and cumbersome and eat some of the power gain. Drives are unlikely on Mars, where every kilogram of cargo is precious and manually adjusting approximately 11,000 m2 of panels (more than 2 football fields) during the day is completely impractical. Therefore, let us assume that our panels will be adjusted seasonally where they point at the sun at solar noon as shown in Figure 16, based on latitude, but not hourly.

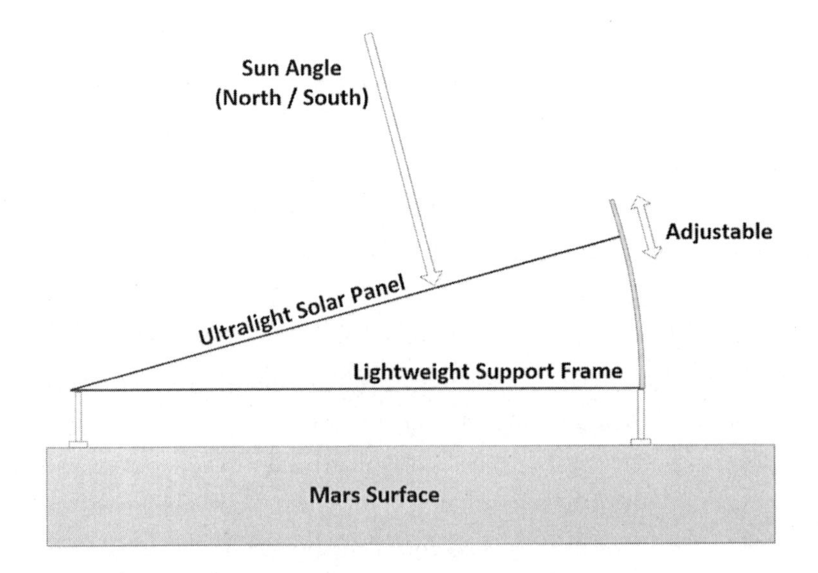

Figure 16: Solar panels pointed at the sun by latitude (and accepting daytime variation)

That means the output is equal to the max times the sine of the angle. Table 8 below demonstrates this.

Time	Sun Angle	W/ m2
6:00 AM	0	0
7:00 AM	15	32
8:00 AM	30	62
9:00 AM	45	88
10:00 AM	60	108
11:00 AM	75	120
12:00 AM	90	124
1:00 PM	75	120
2:00 PM	60	108
3:00 PM	45	88
4:00 PM	30	62
5:00 PM	15	32
6:00 PM	0	0

Table 8: Hourly solar power output per square meter (equinox or on average)

If SpaceX comes up with a very lightweight mechanism for pointing during the day, that only means we are overestimating the number of panels needed, assuming no shadowing (one panel in front of the other). However, any added mass may not be practical.

The area under the curve, and therefore the daily power produced is 948 WH per m2 on average. That is a number we can work with as long as we make light solar panels (which are commercially available), not the heavy glass panels we see on roofs. The backing is not important, just the cells. Something like a Kevlar webbing, similar to a fishing net, could be used to support a flimsy panel. The loading is trivial at 1 kg/m2 or less which is equivalent to 0.38 kg/m2 in Mars's light gravity. It is also possible to trade some solar panel area for Mylar film mirrors to focus more light on the panels, which could help with total mass.

Note that 948 kWH/m2 is per Earth day actually. It is easier to do the math that way. A Mars day is actually about 24 hours and 40 minutes, almost the same, but we are counting the stay in Earth days. As long as margins, like batteries, account for the slightly longer Mars day and night, the result is the same. The actual per Mars day would be 974 kWH/m2 but we would have almost 3% fewer days, which exactly makes up for the power output.

The Sabatier Reactor

The key to producing rocket fuel on mars includes a few simple technologies that are well understood. The first is a Sabatier reactor. We already mentioned it for CO2 recycling in chapter 4. The Sabatier reactor was invented in 1897 by Paul Sabatier and Jean-Baptiste Senderens. It takes CO2 and hydrogen and makes water and methane.

$$CO_2 + 4H_2 \rightarrow CH_4 + 2\ H_2O$$

This is very old technology and used to produce natural gas some places in the world. It is very endothermic. In other words it produces a lot of heat. You just have to preheat it to initiate the reaction. Then it will continue as long as you feed it the ingredients on the left with no additional heat input. In fact you have to cool it after it is up and running. It is most efficiently conducted in a chamber at about 15 bars (~15 atmospheres) of pressure and temperatures around 300–400° C. This means you have an insulated chamber and you preheat it. Other than that, it is easy to use and doesn't even require preheating of gases once it is running because it generates so much heat.

The configuration is basically a tube with a catalyst inside, actually many tubes at the scale we are talking but still the same idea. Each tube is perhaps 2.5 cm (1 inch) in diameter and about 1 meter long (3.3 ft), though almost any size tube works. The main issue is just being able to cool it so it doesn't get too hot and damage the reactor. The most common catalyst to initiate the reaction is nickel because under the right conditions it results in 100% methane production / selection vs. other products being output. It is cheap and effective and studies have shown it to last for a long time, as in months or years under continuous use. Rubidium and rhenium are also options but more expensive. Rubidium though has the advantage of lasting even longer and being more tolerant to multiple shutdowns and startups with little or no degradation. The cost of rubidium is not an issue in the context of a mission to Mars. The quantity involved is very small, as in grams so let's assume we use rubidium as the catalyst.

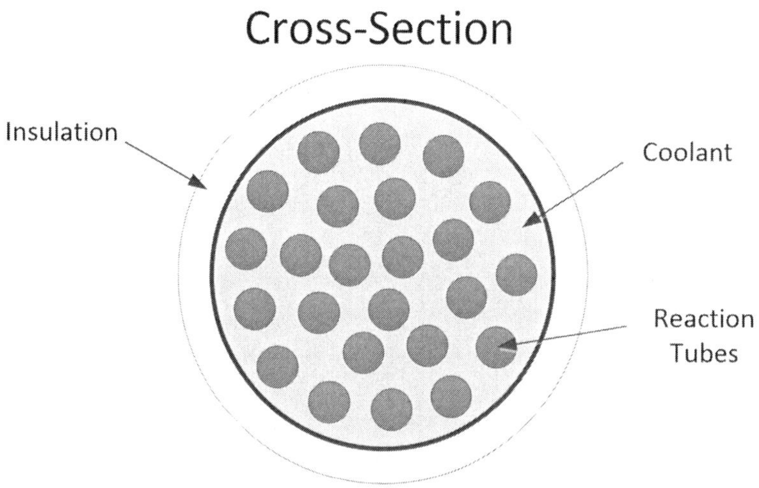

Figure 17: Sabatier Reactor Concept (many tubes surrounded by coolant)

This is just the concept. The actual one would be larger and have more tubes.

Because it generates so much heat, doing this on any significant scale requires cooling of the chamber as shown in Figure17. This can be easily accomplished using a cooling fluid that is stable over the required temperature range of something around 250–350 C. One author considered both molten salts and thermal oil. He found thermal oil to be a superior coolant fluid based on slightly worse thermal properties but much lower power to pump it[9]. It is a liquid at room temperature but stable and liquid up to over 400^0 C. At a relatively low coolant flow he had an average wall temperature of 343^0 C, a maximum hot spot of 623^0 C near the entrance to the reactor where the catalyst starts and most of the reaction takes place, and an average coolant temperature of 268^0 C.

For our purposes, we are going to assume this coolant and a similar flow rate. The flow rate is a trade-off with coolant volume because a fixed amount of heat must be removed to balance the heat input. The idea is just that we have this coolant pulling heat away from the reaction and radiating it to space but we want to keep a stable chamber temperature by limiting coolant flow. Higher coolant flow rates resulted in the walls getting too cool and the reaction not going to as high a percent of completion.

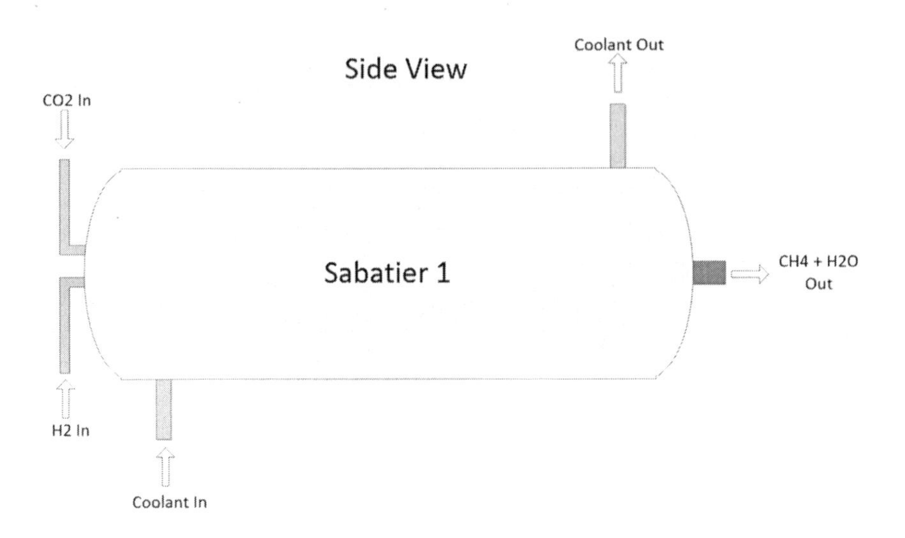

Figure 18: Sabatier Reactor side view

The concept of Figure18 is that CO2 and H2 are fed in from one end. This splits into the many reaction tubes in the end cap. Then the tubes all feed into a common exit tube in the other end cap. It is depicted as white because it has insulation on the outside, with just a hint of transparency to show the size of the actual reactor inside.

Note that at least one researcher demonstrated this technology with excellent efficiency using coolant tubes running down the center of the reaction tube. However, they had issues with temperature gradients and the wall being too hot. (It was a very small-scale laboratory demonstration.) Assuming the catalyst is on the walls, I believe the external cooling as shown in Figure 17 might be more effective. It is similar to commercial units of large capacity, such as those used to produce natural gas. However, when this is built I am sure they will experiment with configurations and select the best one. I am simply doing enough of a concept to estimate mass and power.

One author found that coolant flow direction had no effect on the performance of the reactor[10]. I am showing the coolant moving the same direction as the gas flow. This is because the hot spot is located near the inlet. Most of the reaction occurs as soon as the gases contact the catalyst. For this reason it seemed this was the area we most needed to remove heat, so I made it the inlet. Either option can work but it is likely a baffle near each end may be needed to force even coolant flow across the entire cross-sectional area.

Otherwise there would be hotspots in the corners away from the inlet and outlet.

The only issue is this is not a reaction that goes to completion. It is an equilibrium reaction. Tests in labs have typically produced about 95% to 98%[6][10] conversion with 100% methane selection (produce methane versus something else). This is reassuring but implies we need to recapture the CO2, which condenses out before methane, and recycle it back into the operation. We might also run the mixture just slightly CO2 rich in order to assure all of the hydrogen is reacted. It is a more precious commodity and we can easily recapture the CO2. (That is the opposite of some commercial uses that try to eliminate CO2 in the exhaust entirely.)

One author in reference 9 who worked at Lockheed Martin used a novel hydrogen pump which scavenges the hydrogen gas from the exhaust gases using a 'well-developed solid-polymer electrolyte technology'. I could only find limited material on this technique but I believe they used a slightly hydrogen rich mixture, then recovered the hydrogen with a resulting near 100% conversion. The actual designer of this system may want to consider that option. Apparently either way could work, although the lab was operating on a very small scale and it may be for that reason a chiller to precipitate out CO2 was prohibitive.

In sizing the Sabatier, I used our flowrates, which is only about 12 g/sec (0.42 ounce/sec). A relatively small reactor with less than 100 tubes with an outer diameter of 1" (25.4 mm) and 1/16" (1.5 mm) wall thickness each should be able to handle this. Using stainless steel construction I came up with a total mass of just under 400 kg, and over half of this is the coolant system. I am going to be very conservative and add 50% to this. The shell, for example at 15 bar and a safety factor of 8 only needs to be 0.088" (2.24 mm) thick. This put the entire outer shell including manifolds on both ends at under120 kg. It is 0.38 meter (15 in) diameter by a meter (39 in) long, not including end caps.

I admit operational Earth-bound reactors are not nearly this light, though I have not been able to find full details on any of them, but they don't have to be light. They design it 100 times tougher and heavier than it needs to be with an end that opens and a huge hatch and latching systems in order to service it.

That would not be an optimal space design. The unit should be welded and have no capacity to open it. With a rubidium catalyst we think it will last for many years, perhaps decades with an operational shutdown rate on the order of once per year, due to dust storms or completion of production. Also with the thermal oil there should be zero corrosion of materials, unlike the liquid salt option. It should last forever, but if it ever breaks down, replace it rather than making it 10 times (or 50 times) the weight so you can take it apart. It would be lighter to take a back-up than to go with the heavy option.

In estimating, I assumed stainless steel construction, which is very tolerant to high temperatures, even well above the design temperatures of the reactor. It also has no long-term creep (or rupture) at temperatures like 250–700 C when designed with a safety factor of 8, like I did. This results in a relatively lightweight construction. Over half of the mass is the cooling system actually.

This process is well understood and is in commercial use today. They even run a much smaller one on the ISS to recycle CO2. We are not talking some new technology we have to invent. It has existed for over a hundred years.

The actual Mars capability has been demonstrated in the lab with excellent efficiency, but probably on a much smaller scale. It is very feasible to use the Mars atmosphere and hydrogen from the water on Mars to produce methane and water. Additional references on Sabatier deign.[11][12][13]

Electrolysis

The second important reaction is splitting the water into hydrogen and oxygen. At first, I considered taking hydrogen with you for a significant mass ratio of feeder to fuel production but quickly figured out that is not practical. Hydrogen is extremely low density. Liquid hydrogen is about 72 kg/m3 (4.4 pounds per cubic foot). This is roughly 1/14th the density of water. First of all, it is a challenge just to keep it liquid and not boil it off, since the boiling point is about 20° K (-423 F) though possible with the right vacuum jacketed multi-layer system. However, the incredibly low density means enough of it will not fit in the Starship payload bay even if you had no crew or other cargo. Therefore, we can't take it with us.

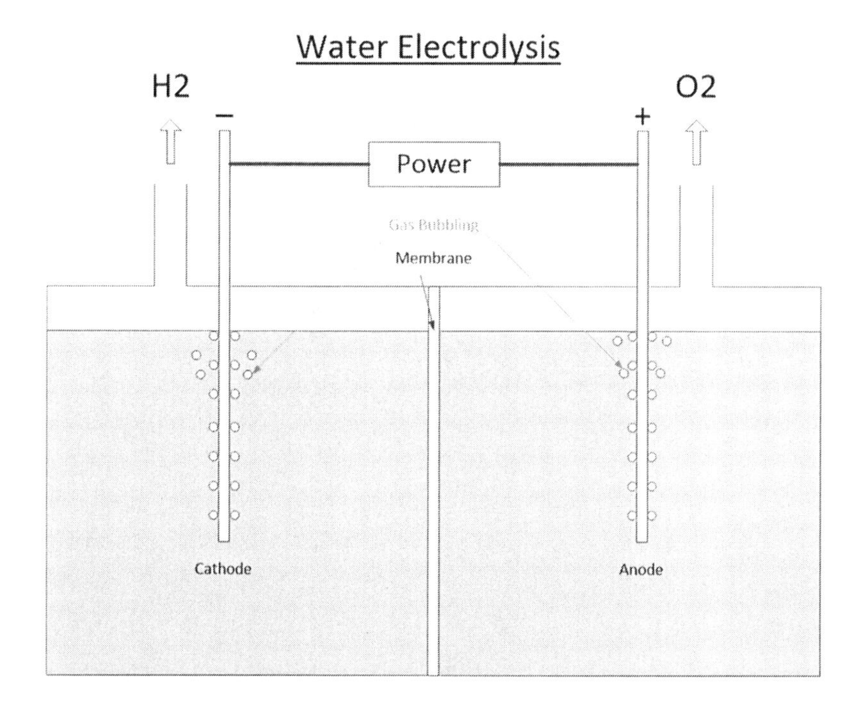

Figure 19: Electrolysis Concept

Instead, we will use electrolysis to create hydrogen and oxygen from water. It is an energy intensive process, the most energy intensive of the entire process, but it is necessary. The reaction is pretty simple. Using electrodes you split water into hydrogen and oxygen. The reverse of this is very energetic and was used to power the Space Shuttle main engines. The issue is we are putting power in to move this in the forward direction.

$$2H_2O \rightarrow 2\ H_2 + O_2$$

This is done in a water tank, preferably at a slightly elevated temperature for energy efficiency. The oxygen bubbles out at the anode and the hydrogen bubbles out at the cathode. You separate these locations and capture the gas. The hydrogen would be pressurized and fed into the Sabatier reactor in order to make methane. The oxygen would be condensed and chilled to fill our oxidizer tank with liquid oxygen.

The concept of electrolysis is shown in Figure 19. It is a water tank and electrodes, along with gas separation and collection. The membrane separates the gases but allows ions to move freely across it in the water. What we will need is to process about 1430 liters of water per 10 or 11 hour day so it will either have multiple tanks or a large tank with large plates for the cathode and anode in order to break down just over 2 liters per minute.

In the past, this reaction was not as efficient as hoped at around 80%. However, recent studies showed a 95–98% efficiency. Their finding was that the bubbles of gas blocked the electric current and interfered with the reaction. By using capillary tubes to remove the bubbles they got the efficiency to 95–98%[14] based on actual power input versus theoretical power input required. That is significant for us when talking about our most energy intensive process.

Our assumption is going to be that we can get 92% efficiency. Given the minimum power requirement per kg of hydrogen is 39.33 kWH/kg and a 92% efficiency the power requirement is about 6,788 kWH per Earth day on average. As stated, this is our most power intensive process, and in fact about 2/3 of our total power usage.

So, the simple fact is if we have water, we can make hydrogen and oxygen. Both are something we need and this is also not high-tech. This is stuff you could buy at home, though I am sure what goes into space will be designed to be optimal for the quantities needed and as light as is practical (probably less than I am estimating).

Note that as shown in the overall system, only half of the hydrogen is converted to methane in the Sabatier reactor. The rest is converted back to water in the reaction, so the water produced is recycled back through the water tank to be separated and go through the process again until it is all converted to methane.

Condensers

In order to use all of the gases, we are producing we need to condense them into liquids. This includes the Sabatier exhaust as well as the O2 from the electrolysis unit. The boiling points of the Sabatier exhaust gases are shown in Table 9.

Commodity	Boiling Point K	Freezing Point K	Notes
H_2O	373	273	
CO_2	233	217	10 bar
Methane	117	91	
Argon	87	84	
N_2	77	63	

Table 9: Boiling points of Sabatier Exhaust Commodities at 1 Bar (1 atm) except as noted

Exhaust from the Sabatier is very hot. The first thing we would do is run it through about a 3 m^2 radiator. This will cool the exhaust to about 400 K before it is introduced into the condenser. Note that it will still be at an elevated pressure and this will be about the point it starts to condense water out. As it is being chilled in the condenser the temperature will be dropped to about 275–278 K. This should condense out over 99.8% of the water, based on the vapor pressure just above freezing. The tiny bit remaining will be captured with the CO_2 and processed back through the Sabatier reactor.

However, CO_2 is an issue. Assuming some restriction to allow time for condensation of the water and the fact that we just condensed out two-thirds of our exhaust gases (water), by this point we are about 2–4 bar (30-60 psia). CO_2 freezes without ever going to a liquid because its triple point is at 5.18 bar as shown in Figure 20. We have to compress the exhaust gases above this to perhaps 10 bar in order to condense it out as a liquid. We then can pass it through a condenser and get out liquid CO_2 to pump back into the CO_2 tank for another pass through the Sabatier reactor.

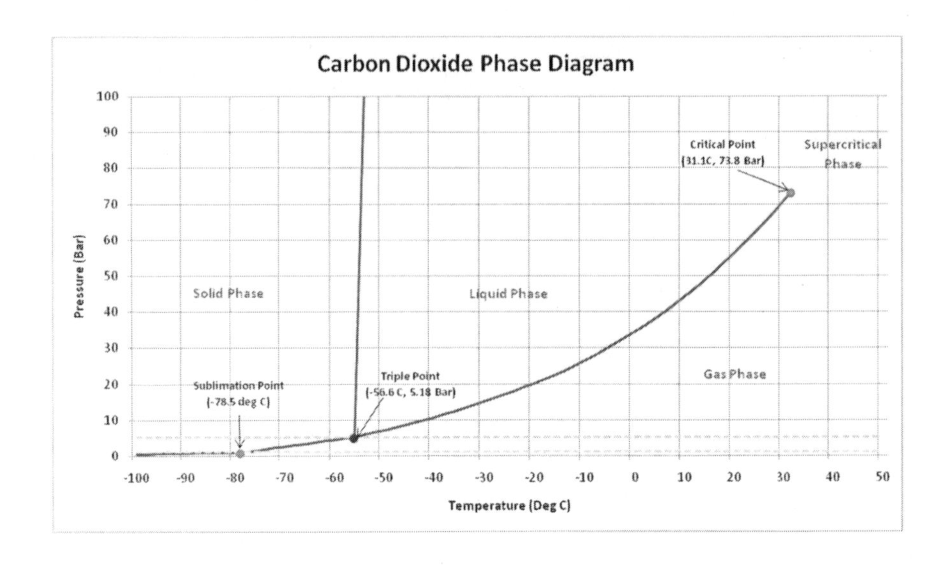

Degrees C

Figure 20: CO2 only exists as a liquid at elevated pressure as shown here

In doing the calculations for power usage of condensers, I calculated the Carnot minimum using the formula:

$$P_{min}= Qx(T_2 - T_1)/ T_1$$

Q is the amount of heat we need to remove. T2 and T1 respectively are the high and low temperatures the cooling system is operating across. This is the minimum number, but no system is 100% efficient. A heat pump, for example is relatively efficient. It can typically remove about 2.4 times the energy it is using. That is why in Florida we use a heat pump to cool and warm our houses. Running the reverse cycle uses only about 42% of the electricity of electric heat.

Cryogenic chillers on the other hand are typically not as efficient. This is because it is very hard to get things that cold.

I tweaked efficiencies until the power usage matched commercially available equipment. In the case of the relatively warm commodities, like condensing out water, I made it match the kind of efficiency you can get from a home air conditioner. This came out as 53% efficiency to match the heat pump. In the case of condensing out the oxygen, I made it match the power usage per kg of liquid oxygen of a commercially available unit that generates liquid oxygen from the atmosphere, after making adjustments for the actual heat removed. The commercially available unit is removing more heat because it is chilling down 78% nitrogen, 21% oxygen. In our case the stream coming out of the electrolysis unit is pure oxygen with some water vapor entrained. My final number was 36% efficiency, or almost three times the power calculated using Carnot's Law.

These efficiencies were then applied to Methane production also, even though it is operating at a slightly higher temperature and would probably be a little more efficient. The results are shown in Table 8, which will follow in a few pages.

In some cases, we may dump the heat to the water tank, which can probably use it, but in most cases we will use radiators. In a near vacuum, like we have on Mars, a radiator can be a pretty good solution. The Space Shuttle and many other spacecraft have used them. There is a formula for heat radiation:

$$\mathbf{P = e \times b \times T^4}$$

P is power in kW (kilowatts)

e is the emissivity, which is a measure of how perfect a material is at emitting energy. It is between zero and one, with one being a perfect emitter.

b is the Steffan-Boltzmann constant, which is 5.6703x10-8 kWm^2/K^4

T is the temperature in degrees Kelvin

For a radiator, this is key. It says if you have a warm plate with your hot coolant in it, you can get rid of a lot of heat by just radiating it into space. Table 10 below was used for calculating the size and therefore mass of radiators. I did not do the detailed multi-node analysis simply because I just want a mass estimate, not a detailed design. The radiators will drive some mass but it is on the order of 900 kg total so +/- 5% is in the noise.

Temp	Avg Surface	W/m2 Out	W/m2 In	Net	Emissivit
220	218	255	101	147	0.95
230	228	305	101	194	0.95
240	238	361	101	247	0.95
250	248	426	101	308	0.95
260	257	498	101	377	0.95
270	267	579	101	454	0.95
280	277	670	101	540	0.95
290	287	770	101	636	0.95
300	297	882	101	742	0.95
310	307	1006	101	860	0.95
320	317	1142	101	989	0.95
330	327	1292	101	1131	0.95
340	337	1456	101	1287	0.95
350	347	1635	101	1457	0.95
360	356	1830	101	1642	0.95
370	366	2042	101	1844	0.95
561	555	10790	101	10154	0.95
551	545	10041	101	9443	0.95
541	536	9331	101	8769	0.95
531	526	8660	101	8131	0.95
521	516	8026	101	7529	0.95

Table 10: Temperature versus watts emitted for heat radiation

Note that this is an estimate, not on the Watts radiated, but on temperatures involved. If we know the temperature in the tube, how hot is the actual outside of the plate and the tube? I estimated it at 99% of the absolute temperature of the coolant in the tube. A rough calculation at 350° K has the center of a two inch aluminum alloy plate between tubes only off 2° K at 348° K and that is the cold spot. This means a 2–5 degree temperature difference between the liquid and the actual radiating surfaces, which is conservative. That is a rough calculation based on the material (probably an aluminum alloy, which is a very good heat conductor if not too hot), a thickness of 1/16 inch (1.6 mm), and distance between tubes of 2 inches (50.8 mm). Also the emissivity can be different depending on materials, but many coatings are available in the range of 0.95 or above, so the value in the table is a reasonable assumption.

Note that the emissivity value is also what fraction it can absorb, at least at that wave length. However, some things are better at one wave length and not another, which is why white paint keeps things cooler than black. Black absorbs visible light, as in sunlight, very well. What we are emitting is primarily infrared. White tends to reflect sunlight and emit well in the infrared.

The reason for the very high temperature numbers in the bottom row of Table 10 is the Sabatier Reactor coolant is very hot and will be in this range. It was estimated using the last 5 rows for each 20% of the heat over about a 50 degree range.

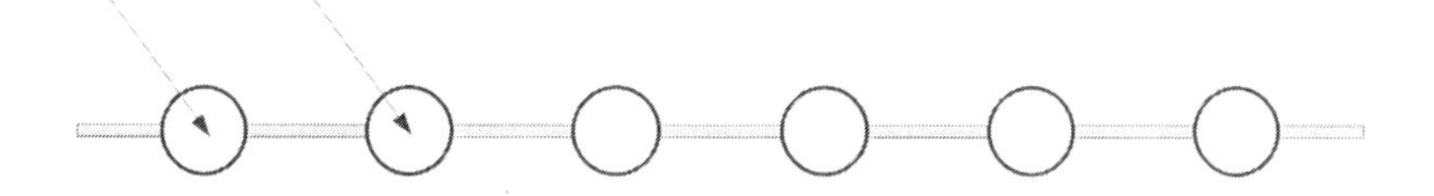

Figure 21: Example of a possible radiator configuration

This table also has an assumption on radiating heat. It is emitting at a temperature slightly less than the fluid inside. It is also receiving heat on its surface. If the radiator is not in the sun it is receiving the natural temperature of Mars, which is pretty cold, on one side and something less on the top side. A way of accomplishing this reduced heat input is to put it in the shade behind the ship, though we don't want to heat the ship up, or at least not the fuel tanks. Another strategy is to turn it where the edge is toward the sun. Note that the plates between tubes in Figure 21 increase the area but at the cost of being slightly cooler. This is part of why the temperature is not that straight-forward. It is not uniform and the input, which reduces how much we can radiate, is a function of orientation and proximity to anything warm.

Compressors

We need at least two relatively large compressors and perhaps two small ones. The largest is to compress atmosphere for use in the Sabatier. The other moderately large one is to compress Hydrogen generated through electrolysis. Both are gases, the first at about 7 mbar; the second at about 1,000 mbar (1 bar or about 1 atmosphere). However, we are going to compress the CO2 to 40 bar and the hydrogen to 120 bar in order to feed the Sabatier Reactor. The hydrogen is a higher pressure only because it is such low density. We are trading a little bit of power for mass (and size) in the tank volume. The concept is that we pressurize them in tanks and use a regulator flow control system to feed them into the reactor. In order to do this, we need tanks, compressors, and a feed system.

Compressors are relatively low-tech equipment. There is actually a calculator online that predicts the power consumption. It is located at https://checalc.com/calc/compress.html. This was used for the calculations. You enter the starting pressure, ending pressure, flow rate in, and a lot of info about the critical point of the gas involved. It has defaults on the efficiency that can be tweaked, though I didn't. They included polytropic efficiency (98%), mechanical efficiency (80%), and interstage pressure drop (0.35 Bar). Based on these, I used its results for the power required. Those results are included in tables that will follow.

Regolith

It is highly unlikely that we will have clean ice for use. The best case is we are collecting it in front end loaders and it is mostly pure ice that is dirty. The more likely case is that we are using a regolith / ice mix that has a high water (ice) content. At least one rover found regolith that was 60% water ice by weight just centimeters under the surface at 60 degrees latitude. Maps that will follow show that this is the norm in some areas of Mars. For the purpose of calculation I assumed 50%. The reason this is important is that to boil it off we have to heat the regolith and the ice. That means a lot of heat input in order to take frozen regolith, melt it, and then heat it up hot enough to boil off the water so that we can capture it and condense it like a distillery does with alcohol. The percent water is important as we will discuss later.

The concept of doing this might look something like Figure 22.

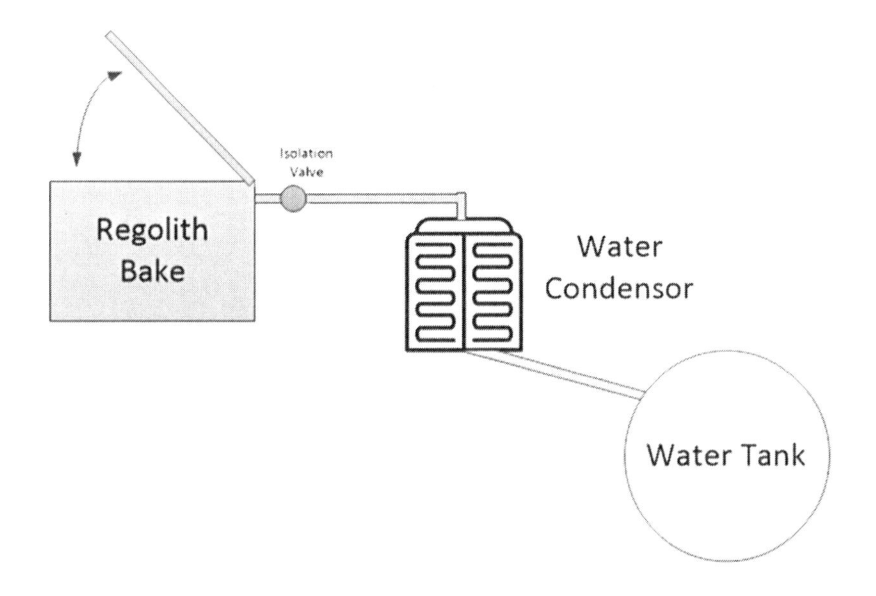

Figure 22: The Regolith water recapture system concept

This operation would have to be performed in batches. The reason is that we need to put the regolith into a microwave oven using a front end loader or backhoe. We will find that we need almost a ton and a half per day, in order to produce 715 kg of water. In order to do this, it needs a lid and an isolation valve. The lid is in order to load it. The isolation valve is so we don't open up our water tank to the atmosphere and boil off our precious water. It also needs the capability to remove the spent, now dry, regolith. This might be as simple as a trap door to dump it on the ground, where the front-end loader scoops it away and adds it to a spent regolith pile.

Some minor additions might be worthwhile, such as a small pump to pump any water vapor out of the regolith bake tank and into the condenser system before opening the regolith bake up to the 7 mbar atmosphere. We also might add a check-valve between the regolith bake and the condenser just to assure no back-flow in the system.

The concept is that once we close the lid and seal it, we turn on the regolith-bake microwave. It first melts it, then heats it further and boils the water. The water vapor escapes and goes into the condenser. The condenser has a chiller cooling it down to a few degrees above freezing. The result is the water condenses and drains into a large water tank. This water tank is used to feed the electrolysis system.

I wish I could say we are just heating the water but we aren't. Ice does not capture microwaves as well as water, so initially we are heating the regolith and the ice with it in order to melt the ice. Once it melts, water absorbs microwaves very well. In fact exciting the water molecules was the entire concept of the microwave and the reason for the choice of frequency. The water will boil readily as it is heated. However, I assumed we are heating the regolith too, which is a bit conservative, but we are passing steam through it. A big part of the energy though is the melting and then boiling of the water. The energy to do those is about 74% of the total energy used. Only 26% is heating the water and regolith between its starting temperature and the melting and boiling points of water.

The quality of the regolith is important. By that I mean the percent ice. To distill water out of regolith like described here uses about 15% of its energy for heating the regolith. If the soil were only 20% ice the power requirement would go up by 130 kWH per day. If it were only 10% it would go up by another 210 kWH per day. Also, the amount of regolith we have to load goes up by a factor of 2.5 and then doubles again in this scenario to 20% and then 10%. That would mean having to have 5 times the capacity volume wise, not just the addition of hundreds of square meters of solar panels. We really need to have ice-rich soil to make this manageable. We can't boil over 7 tons a day to get the water we need, and both power and volume go asymptotically toward infinity as we approach zero water content. This is shown in Figure 23.

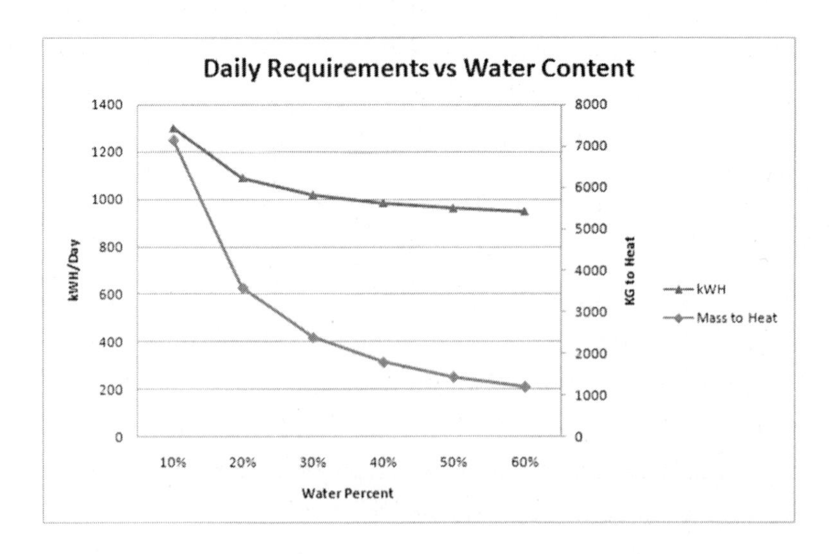

Figure 23: Power and mass of regolith daily versus water content

The power required to boil off water from the regolith is shown in Table 11. This is the actual power emitted by the microwave.

H2O Boiloff							
			kg/hr	Kstart	Kfinish	kWh/Wh	kW
Icy Soil	1.585	KJ/kg K	155.00	188	273	0.000278	5.80
H2O	336	KJ/kg	91.14			0.000278	8.51
Soil	0.800	KJ/kg K	63.86			0.000278	1.89
H2O	4.187	KJ/kg K	91.14	273	373	0.000278	14.4
H2O	2260	KJ/kg	91.14	273	373	0.000278	57.26
Total							**87.62**

Table11: Power required to boil water out of regolith. This assumes only 8 hours per day.

Note that this is the actual power imparted to the materials. A microwave would be about 85% efficient so the actual power required would be over 100 kW (plus the condenser).

It might be possible to simply liquefy the water and filter it as it drains off, avoiding both boiling and condensation. This occurred to me but how you clean the filtration system which would be covered in regolith and actually get pure water for the electrolysis on this scale is not apparent to me so I did not assume this. There are also issues with lots of dissolved salts from the soil winding up in the electrolysis equipment. I assumed the more power intensive task of boiling it off and then condensing it. Perhaps someone more clever can design such a system that could filter out dissolved salts, not just the dirt. The system above was assumed and power allocated accordingly.

Scaling It Up

We have discussed the basic components of the system and even shown an overall schematic. What we haven't really done is talk about the scale of the system as a whole. How much of each thing do we need and what is its mass?

First, let's discuss how much propellant we need.

The first chemical formula is for the Sabatier system:

$$C_2O + 4H_2 \rightarrow CH_4 + 2H_2O$$

This says that for every molecule of methane we need one molecule of carbon dioxide and four molecules of hydrogen. We also produce two water molecules.

The other main reaction of interest is electrolysis:

$$\mathbf{2H_2O \rightarrow 2H_2 + O_2}$$

This says we can take 2 molecules of water and produce the hydrogen we need plus oxygen. However, we have to do twice as much electrolysis to feed the Sabatier process. We also create water in the Sabatier process so half of our water will come from recapturing that water and only half from the regolith. In essence half of our hydrogen molecules go right back to form water. We keep recapturing that and passing it through the Sabatier process again.

This means the ratios are:

1 CO_2 to 1 CH_4
4 H_2 to 1 $C H_4$
2 O_2 to 1 $C H_4$

However, an O2 molecule has an atomic mass of 32 and CH4 has an atomic mass of 16. Therefore the 2 to 1 is really 4 to 1 by mass. Since our mixture ratio (oxygen to methane) is 3.8 this means we are producing a slight excess of oxygen. That is not a bad thing on a planet with very little of it.

The quantities are summarized in Table12 below.

Propellant Requirements			
US Mass:		120 tonnes	
Return Payload		4 tonnes	
MR for Earth Ret		6.82	
Margin:		10 tonnes	
Total		731.68 tonnes	
Propellant Quantities:			
Target:	480	Days	
	Total (tonnes)	Per Day (tonnes)	
Needed	732	1.52	
M/R	3.8		
Methane	152	0.318	
O2	579	1.207	
	Mass	Number	kg/day
CO2	44	1	873.32
H2	2	4	158.78
CH4	16	1	317.57
O2	32	2	1270.28
H2O	18	4	1429.06

Table 12: Propellant and supporting commodity mass requirements

The top area is a summary of what we need for the flight back to Earth. The upper portion of the table with borders is what we need based on mixture ratio. The bottom portion, which includes kg per day is what we will produce or need to get the required amount of methane and oxygen. One thing to note is what we will produce includes a slight excess of oxygen by about 63 kg/day (compare bottom area to top area). This is not a bad problem. In making fuel we create an excess of oxygen that can be used to provide portable oxygen, such as for use on a rover, or can be used to provide atmosphere to habitats and greenhouses.

These numbers are pretty large because we have a big rocket. We need to produce 732 tons of fuel for its return to Earth. To do that is 480 days, based on a standard 550 day stay less some margins for startup and possible lost time due to dust storms, means about a ton and a half per day. That is a challenge, but not insurmountable. From there we do the math and figure out what we need. Note that I am ignoring the few tons we landed with as margin. I am going to be a little conservative and assume that makes up for boil-off and other inefficiencies.

We need 873 kg per day of CO_2 from the atmosphere. Since the atmosphere is 95.32% CO_2, we have to compress about 916 kg of atmosphere per day. That is very possible to do, but it means a pretty large multi-stage compressor. Using the online calculator at https://checalc.com/calc/compress.html and using 10 hours per day I get 86 kW (115 HP). I considered 24 hour and lower power, but it is pretty linear and added more weight for batteries than the added compressor weight.

We also need 1429 kg of water every day. Fortunately, half of this is recaptured from the Sabatier exhaust, consistent with our previous discussion that half of the hydrogen is converted back to water. Therefore we need 715 kg from the regolith. If it is 50% water by weight we need 1430 kg of regolith per day. That much regolith is not something that will be done with a shovel. My thought is an electrical front-end loader and/or a backhoe that can scoop the soil up and dump it into the boil-off chamber.

The O_2 and H_2 are only internal products derived from water. Their rates are important but they are generated, not directly removed from the environment.

The Sabatier process is best left to run 24/7. It takes some power to start it up. You have to get the chamber up to 300–400º C. However, once you start it, it will keep running on its own because it is very exothermal. In fact, you have to provide cooling in order to not have other issues. That does not mean the power requirements are zero. You have to run the water, CO_2, and methane condensers as well as the Sabatier cooling system.

The power requirement of these condensers was calculated with an efficiency factor between 36% and 53% as described earlier in this chapter. These condensers are chillers, but they must dump the heat being removed somewhere, just like the outside unit on your home AC. This is assumed to be radiators and the size and mass was calculated for each.

The electrolysis process is way too power intensive to run 24/7, including on batteries at night. The intent is to run it about 10 to 11 hours per day (plus or minus depending on season) while solar power output is significant. Otherwise the battery requirement would be massive, not to mention minor losses due to storing the energy and then trying to get it back.

The processes require:

- Electrolysis to split hydrogen and oxygen

- Atmospheric compression
- Hydrogen compressor to pressurize hydrogen for Sabatier usage
- The Sabatier and its cooling
- Condensation of water, CO2, and methane from the Sabatier output
- Pumping methane to the vehicle
- Condensation of water from the oxygen stream from electrolysis
- Condensation of oxygen and chilling it for transfer to the vehicle
- Pumping LOX to the vehicle

These processes are relatively power intensive, but have each been looked at. The final outcome is shown in Table13 below.

Power Budget:					
		Daily Budget	**Calc energy**	**Efficiency**	**Hrs/Day**
Sabatier:	Sab cool Pump	9	8.94	70%	24
	H20 Condenser	199	105.48	53%	24
	Exhaust Comp	6	6.12	78%	24
	CO2 Condenser	10	4.14	40%	24
	CH4 Condenser	338	121.82	36%	24
Electrolysis & O	Electrolysis	6788	39.33	92%	10
	H2 Compressor	249	24.87	78%	10
	02 Comp(Extra)	22	2.18	78%	10
	H2O Condenser	13	7.14	53%	10
	02 Condenser	918	330.58	36%	10
Support:	Atm Compressor	860	86.00	78%	10
	Regolith Boiloff	948			8
	Pumps & Margin	25			
Total:		**10385**			

Table 13: Power requirements for power production

This says we need over 10,000 kWH per day before considering charging vehicle batteries and consumption to support humans. This is not trivial but can be accomplished.

So, let's look at power. In order to produce this amount plus some margin to run the ship and charge electrical vehicles let's add in some margin and say we need 11,400 m2 of solar panels. This power output looks something like Table14.

Solar Panel Power Output			
Area		11400 m2	
Assuming stationary panels pointed at the sun latitude-wise			
Time	**Angle**	**Power/m2**	**Output kW**
6:00 AM	0	0	0
7:00 AM	15	32	367
8:00 AM	30	62	708
9:00 AM	45	88	1002
10:00 AM	60	108	1227
11:00 AM	75	120	1368
12:00 AM	90	124	1416
1:00 PM	75	120	1368
2:00 PM	60	108	1227
3:00 PM	45	88	1002
4:00 PM	30	62	708
5:00 PM	15	32	367
6:00 PM	0	0	0
* Average day, as in equinox			
Total:		**10,821** kWHr/day	

Table14: Solar power output

A further consideration is how we manage that power on a daily basis. We said we are only going to run electrolysis during daylight hours due to the huge power consumption and lack of need to run it continuously. We provided hydrogen tanks as a daily buffer to allow an asynchronous production from the Sabatier and electrolysis. We further need to manage power on an hourly basis. Table 15 is a suggested plan, where we ramp up electrolysis production during peak hours. The Sabatier runs at a constant rate 24/7 but the really power intensive stuff we only do during daylight hours. Note that I am assuming half an hour after sunrise to half an hour before sunset. We probably could run a little longer, but it would be at very low power until the sun angle finally gets high enough to produce significant power.

Daily Power Management								
Time	Outp ut kW	Elecroly sis	O2 & H2	Sabtier Sup	Regoli th	Atm Com p	Tota l	Margi n
6:30AM	185	90	16	24	0	0	130	55
7:30AM	542	310	56	24	0	86	475	67
8:30AM	862	487	87	24	118	86	802	60
9:30AM	1124	706	127	24	118	86	1060	63
10:30AM	1309	861	154	24	118	86	1243	66
11:30AM	1404	941	169	24	118	86	1337	67
12:30PM	1404	941	169	24	118	86	1337	67
1:30PM	1309	861	154	24	118	86	1243	66
2:30PM	1124	706	127	24	118	86	1060	63
3:30PM	862	487	87	24	118	86	802	60
4:30PM	542	310	56	24	0	86	475	67
5:30PM	185	90	16	24	0	0	130	55

Table 15: Daily power management

As this shows, power usage needs to be balanced with power production during the day. Solar power turns on and off during the day/night cycle and must be managed. Note that this shows an excess of about 65 kW during the day. This allows charging batteries, running the spacecraft, and charging heavy equipment and the rover. At night power usage would probably be minimized, though the intent is to run the Sabatier reactor all night and heat might be more needed at night. There is an allowance in the mass for seven 75 kWH batteries in order to provide up to 525 kWH of power during the night. This is about 14 hours of Sabatier usage plus a 50% margin above this for heating and running the ship during the night.

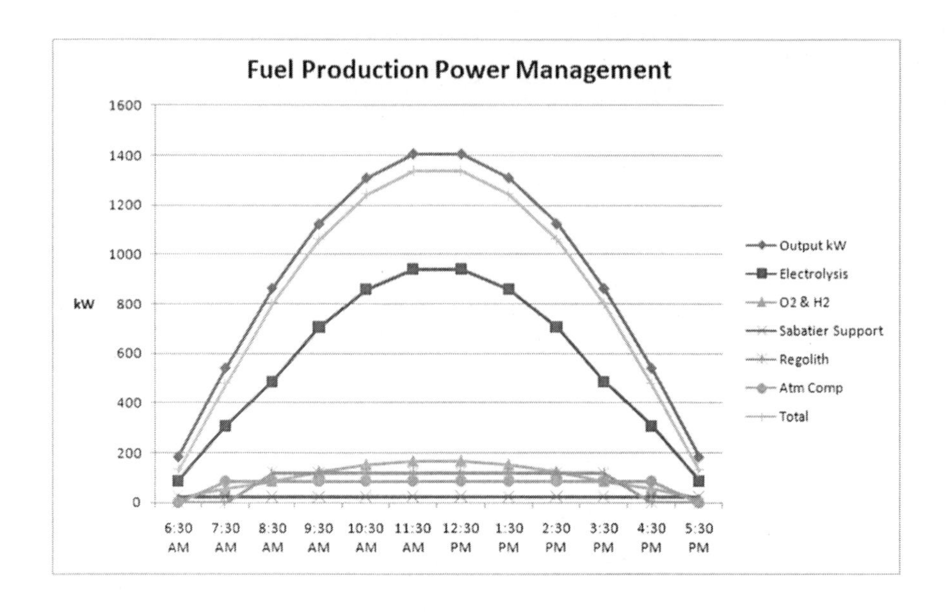

Figure 24: Daily power budget

The daily power budget is shown graphically in Figure 24.

Our daily power production is about 436 kWH above that required for the fuel production. If we assume four vehicles with batteries and we have to do a 75% charge daily that is just over half of this. That leaves about 211 kWH or about 8.8kW around the clock to run the ship or other miscellaneous uses.

Now let's look at masses. Solar panels will be our power and we need a lot of them. They produce about 954 WH per square meter per day on average. As discussed previously, the panels will be very light and supported by a very light frame that allows angular adjustment. I am estimating 3 kg per m2 total. There are actually commercially available panels at 1.5 kg/m2 so I don't think this is too out of line if a light enough frame can be designed that isn't so flimsy it gets damaged just trying to set it up.

Table 16 summarizes the masses for power production.

Fuel Production Mass Requirements:			
Item	Quantity	Mass Per (kg each)	Total (tonnes)
Solar Panels	11400	1	11.40
Solar Frames	2850	8	22.80
CO2 Tank	1	126	0.13
H2 Tank	4	544	2.18
02 Tank	8	358	2.86
Water Tank	1	137	0.14
Sabatier	1	600	0.60
Electrolysis	1	300	0.30
Atm Compr	1	1100	1.10
H2 Compr	1	300	0.30
02 Compr	1	40	0.04
Liquid Pumps	3	30	0.09
Cryo Pumps	2	100	0.20
H2O Consenser (Sab)	1	250	0.25
CO2 Condenser(Sab)	1	100	0.10
CH4 Condenser	1	400	0.40
H20 Condenser(O2)	1	100	0.10
02 Condenser	1	600	0.60
Plumbing	1	80	0.08
Radiators	65	13.5	0.88
Batteries	7	480	3.36
Umbilicals	4	100	0.40
Regolith Boils	1	2175	2.17
Structure and Margin	1	1000	1.00
		Total:	**51.5**

Table 16: Mass requirements for fuel production

The H20, H2, O2, and CO2 tank masses were calculated based on the volume and pressure required assuming a titanium alloy with a safety factor of 1.5 to 40 depending on the application and expected other stresses. These tanks act as a buffer to production. Hydrogen is only produced during the day but is used 24 hours a day. Therefore we need tanks for a buffer overnight. The other tanks also act as buffers in the process allowing continuous usage regardless of fluctuations in the leading step. This process as described is very tolerant to the daily asynchronicity of the four processes, the Sabatier, atmospheric compression, water production, and electrolysis. The oxygen tanks are not really part of the process but are intended to capture the excess oxygen we don't need but could use elsewhere.

Overall, I believe this estimate is pretty close and being even five percent high or low does not change the viability of the plan, just a few details, and I think I have erred on the conservative side (over-estimating mass) in most cases.

Combined, all of these numbers give us a basis for mission planning. How much of our payload is eaten up by fuel production? Perhaps this is a lot of detail, but a lot of detail goes into real planning. It is my intent to arrive at a rationale estimate of what we can and cannot do, or perhaps I should say what SpaceX can and cannot do.

7. Where We Will Go

I shall be telling this with a sigh
Somewhere ages and ages hence:
Two roads diverged in a wood, and I—
I took the one less traveled by,
And that has made all the difference.

Robert Frost, Poet

The ideal location for settlement is near the equator. The temperature is higher. The sun light is better. It is much better for growing crops and for limiting the need for heating everything. However, a mapping of mars shows very little water there and we need water. Look at Figures 25 and 26. Figure 25 is very revealing in that it shows the water content is in high latitudes, near the poles. This is NASA's 'treasure map' of where the water ice is. The white box is their prime landing area around Arcadia Planitia. It is an area rich in water ice almost right at the surface but not at extreme latitudes (not near the poles). The scale is the depth to get to water ice in fractions of a meter.

Figure 26 shows the actual content in the first meter by percent. We need high water content to extract water. While we can take heavy equipment to Mars, well smaller electric ones, in order to dig up ice we are probably dealing with regolith (soil) that is high in water ice content, at least on the early missions. The energy penalty is severe for trying to remove water from regolith that is only single digits in water ice content. We need the higher levels.

The reason water content needs to be high is to extract the water we need to thaw it, then boil it off, and then condense it. This means raising the temperature to 100 C from somewhere well below zero. We don't want to do that to tons of regolith in order to just get a few kg of water. We want a mixture that is closer to 50/50 or even above in order to manage the power levels and volume required.

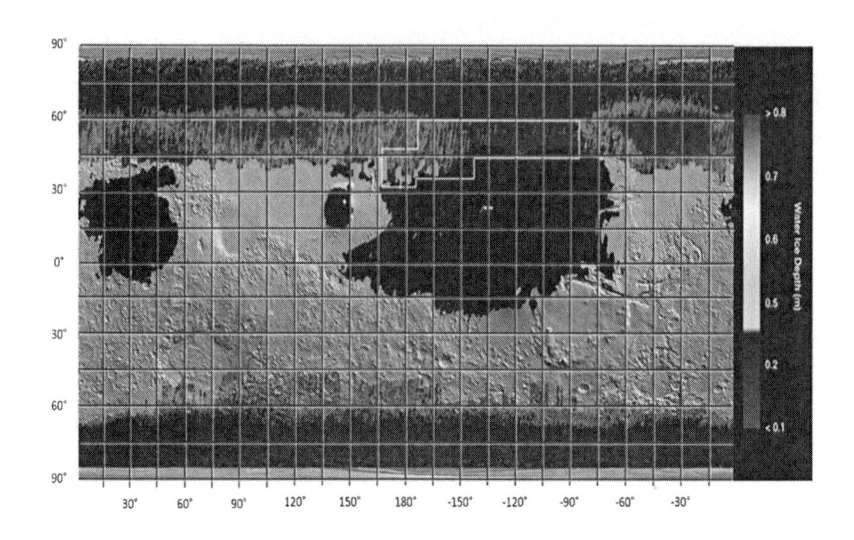

Figure 25: Current NASA map of water on Mars with latitude and longitude
Credits: NASA (except for the grid and latitude and longitude labels)

Will we find water at the equator? In 2018 NASA's InSight lander touched down in an equatorial region in an area known as Elysium Planitia. It conducted seismic tests to gain insight into what lay beneath. The scientists' conclusion from the data was there was no appreciable water within the first 300 meters[15] (985 ft). That means that at the equator we might not find water even if we dig or drill. This is pretty consistent with both maps. Figure 25 shows no ice within at least the range of their scale, up to a meter. Figure 26 shows some water content near the equator but in the single digits percentage-wise at best. As we said before, that is a power issue and volume issue. It might be possible to extract some water but probably not on the scale we are talking about. We need almost a ton per day.

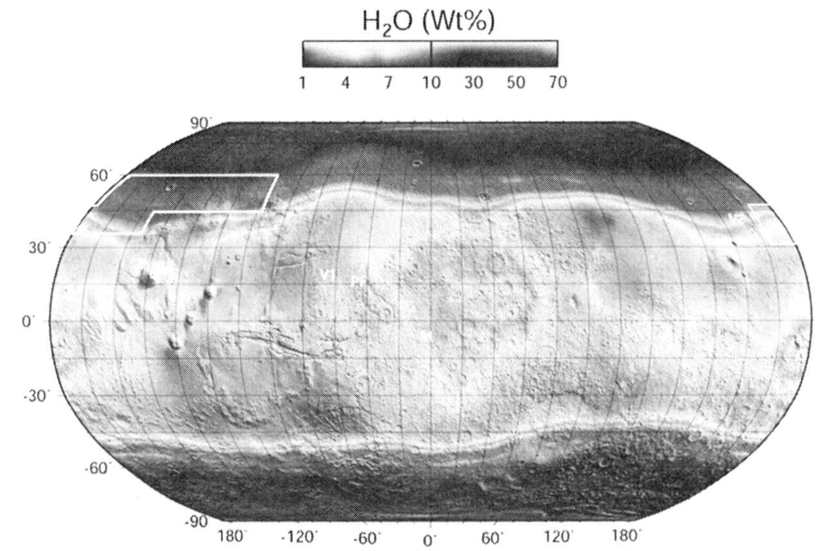

Figure 26: Odyssey map of ice content in the first meter of regolith
Credits: NASA

Note that in Figure 25, they did not provide any latitude or longitude data. By picking out features like the Valles Marineris, the giant canyon on Mars that is just south of the equator between about minus 60 and minus 90 degrees longitude I was able to decipher what they did and add the grid lines. Figure 25 stretches out the edges to get a rectangular perspective. They also shifted the view 180 degrees relative to Figure 26. The left half of Figure 26 is the right half of figure 25. In other words they put 0 degrees latitude on the edges instead of in the middle.

Another point is that black areas in Figure 25, NASA claims, are fine dust that a landing craft could sink into. I recall they were very afraid of that on the moon too, which may be part of why the lunar lander had very broad landing pads. It turned out it was very packed fine powder, which meant a good soft landing surface like landing on a grassy field. My belief is it is always possible to design around issues like that but it is something to be aware of.

Also note that I added NASA's prime landing target Arcadia Planitia box onto Figure 26. It looks slightly distorted because of the perspective.

Why Not the Poles

The poles are perfect. They have ice right at the surface and it is nearly pure water. Just what we are looking for, right? Well not exactly. The poles might be rich in resources but they are in the dark and extremely cold for the half a year where they do not face the sun. Just like our poles, everything in the Arctic Circle is dark for half the year and then light for half the year. That is because Mars is on an axis of 25 degrees, similar to the Earth's 23 degree axis, which drives winter and summer. For us that means 343 Earth days with no solar power. We can't survive that. We couldn't carry enough batteries to keep astronauts alive for that long.

The poles are not an option for a solar power-based mission.

The Right Compromise

If my life was on the line, there is no way I would go to the equator on the first mission. We are very likely not to find water in quantities we can use. On the other hand, there is no way I would go to the poles, where water is plentiful but I will freeze to death in the winter. So, what is the right answer?

There are places in between with lots of water and year round sunlight. For our purposes we will choose 45 degrees latitude. If we decide we can land at Arcadia Planitia in NASA's prime target box and be able to land at the same longitude much closer to the equator later we may be able to even go another 5 degrees south and have water ice within a meter of the surface. If we decide against this option, we can still go to between 90 degrees and 120 degrees longitude at 45 degrees latitude and find water ice near the surface with plenty of low terrain non-mountainous areas south of us.

So, let's say the first mission or two go to 45 degrees latitude. Where does that leave us?

Well, first of all, we can probably dig up all the regolith we want that is 50% or greater water content by weight and it is within about 0.1 to 0.5 meters (4–20 in) of the surface. Understand Mars once had oceans that probably froze and testing has shown this level or close to it at the surface at this latitude. It is probably a lot wetter below, since water tends to sublimate at the surface (go from a solid to a gas). We might even hit pure ice. Since we intend to take heavy equipment, well small ones that run on batteries, moving dirt is not an issue. We could easily scrape off the first half a meter and use all of the ice rich regolith we want. Note that the Phoenix Rover scraped off about 5 centimeters and found water at 60% at a little above this latitude. We will have ice, which makes water, which fuels our electrolysis to make the whole thing work.

So how do we do that? We need solar power. The atmosphere is very thin and solar power should be close to the same as long as we can point the panels at the sun. This presents a bit of a problem but is actually easy to solve. What we need is to be close to a slope, like the side of a crater or mountain. If we have a 15–20 degrees slope we can line up our solar panels on the slope so we can turn them at a high angle but not have shadowing. This is shown in Figure 27 with the worst-case winter solstice angle.

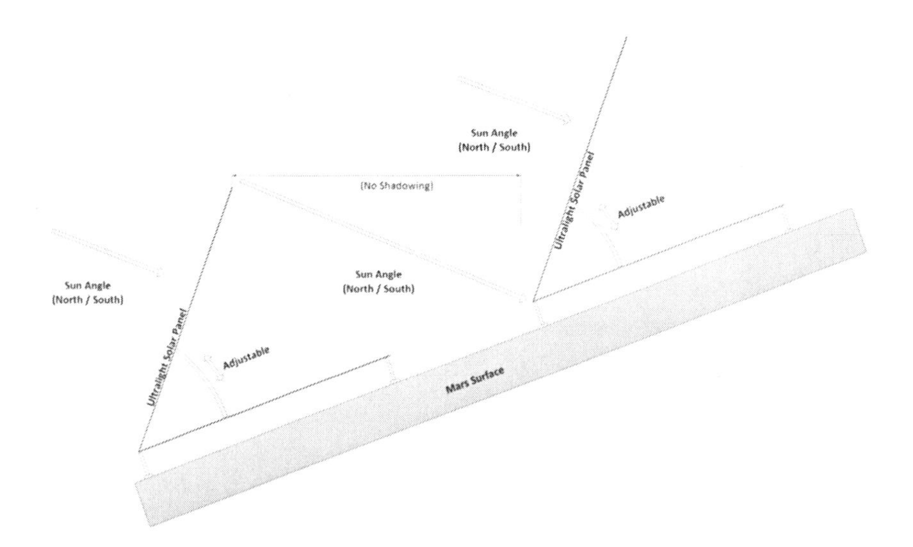

Figure 27: At a high latitude, arranging solar panels on a slope minimizes distance required to prevent shadowing in the winter

This slope of 20 degrees reduces the distance for no shadowing from about 3 times the height of the panel to about 1.5 times the height of the panel, so this is how we manage power production if we can. If we can't find a slope we just spread them out farther, but we are already at the size of a couple of football fields.

There are also issues with longer day/night cycles in the winter at northern latitudes. This drives a little bit of power management and sizing. The Sabatier process has to keep up with the electrolysis, which is running about 14 hours a day at the extreme of summer (summer solstice) and only about 6.7 hours at the extreme of winter (winter solstice). This is daylight hours minus 2 since solar power is extremely low in the first and last hour of daylight with us not having a mechanism to rotate them east or west. This means higher production when power is more available. However, it does provide a colder background for heat rejection, meaning chillers can be a little more efficient. Otherwise, the higher latitude is not much of a challenge.

Note also that the northern hemisphere tends to have lower elevations. This means better aero-braking for vehicles coming in. It also means slightly warmer, slightly more atmosphere for fuel production, and lower radiation levels. Therefore, let's assume this is 45 degrees north latitude.

Where We Want to Be

Where we really want to be is in a tropical region. The sun is much better for growing crops. The days are a little warmer, so it reduces heating requirements. We also don't have those long winter nights that drive battery usage.

Let's assume we choose our ideal spot at 15 degrees latitude north. This is well within the tropics and near the equator. It is also a little bit closer to our water depot at 45 degrees than more southerly locations. Hopefully we chose wisely on our first location so these two locations are approximately due north-south from each other. This will be our primary location for colonization. Fifteen degrees was chosen over a lower latitude because with the elliptical orbit this actually provides a more consistent solar flux level.

On later missions, we will look for water sources here. It is possible that there is an underground river nearby or better yet a warm geothermal stream which could provide water and power. This will require scientific testing, probably seismic or radar, in order to find a probable location to drill. We might have to drill as deep as a kilometer or more, but if we do find water we have the choice of shutting down our water station at 45 degrees and moving everyone here.

The benefit of this approach is we may become water independent eventually at our ideal location, but we never risk lives by sending them to a possibly arid location with no water available to get them back.

Where Does This Leave Us

Perhaps we should look at the maps again to make sure we have chosen wisely. On each map, I have marked the approximate location of our possible site with a white X, showing both the northern and tropical locations. First, let's look at our water map, which includes terrain, except where it is blacked out. This is shown in Figure 28.

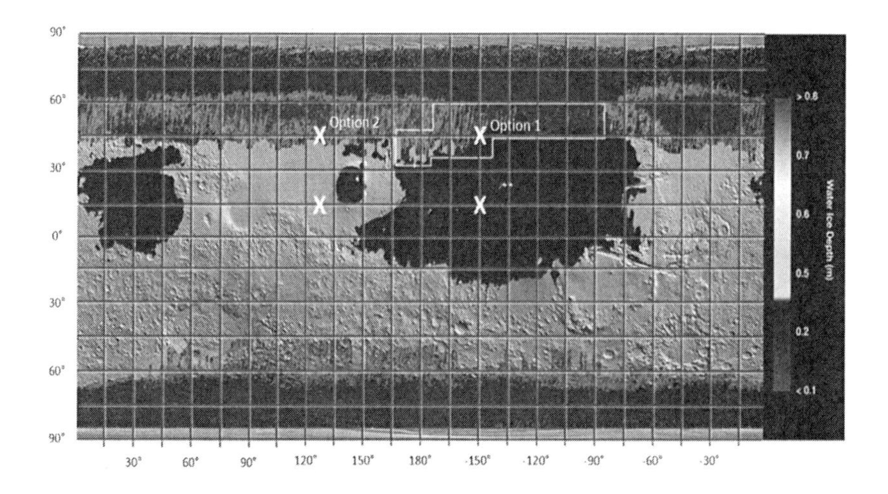

Figure 28: Water map with locations of bases
Credits: NASA

These locations look reasonable with regards to water and terrain except where the terrain is blacked out on the Option 1 route between locations. As discussed previously, Option 1 requires perhaps enlarging landing pads to account for the softer surface. However, one other possible thought on the 'fine powder' is that what a scientist sees as fine powder a farmer sees as perfect soil for growing things once we add some nitrogen. I assume NASA is basing this on grain size. Good farming soil does not contain rocks. It contains granular dirt. Perhaps option 1 is the better choice for a permanent base. Our Mars inhabitants will be growing crops in a few years at the tropical location. They need native soil that is not too rocky. Travel between sites is also much easier, or at least less bumpy, without a rocky terrain. As you could see back in the picture of Curiosity Mars' surface can be quite rocky in places.

The next map is the surface water map that shows the percent water in the first meter. This is shown in Figure 29.

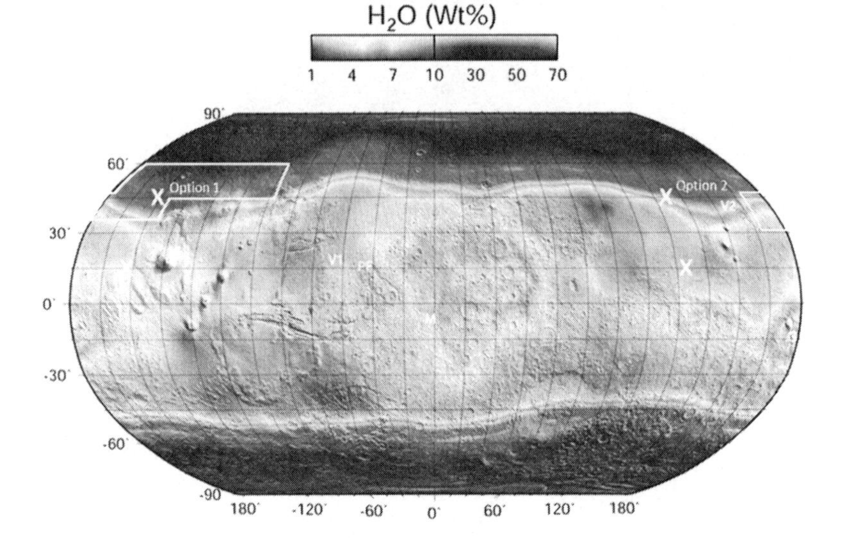

Figure 29: Surface water map with landing locations shown
Credits: NASA

This appears acceptable as well. Both options provide a high water content near the surface at 45 degrees latitude. This is because the ice is less than a meter down, as shown in the previous figure. At the tropical location both are in the 3 to 4 percent range at the surface but have no indication of ice near the surface in the previous figure. This is acceptable because we are not going to rely on them to make their own water, but they could at a much slower rate. Either tropical site may have water deep underground but we don't know that until we get there and run tests looking for it.

Figure 30 shows where past Mars spacecraft have landed. Phoenix found 60% water within centimeters of the surface. InSight found no indication of significant water within 300 meters (seismic testing). However, I believe InSight was looking for underground rivers, not just ice encrusted into the regolith. Our stated options are marked on the map with 'X'. These are places we could land for a first mission and follow on missions.

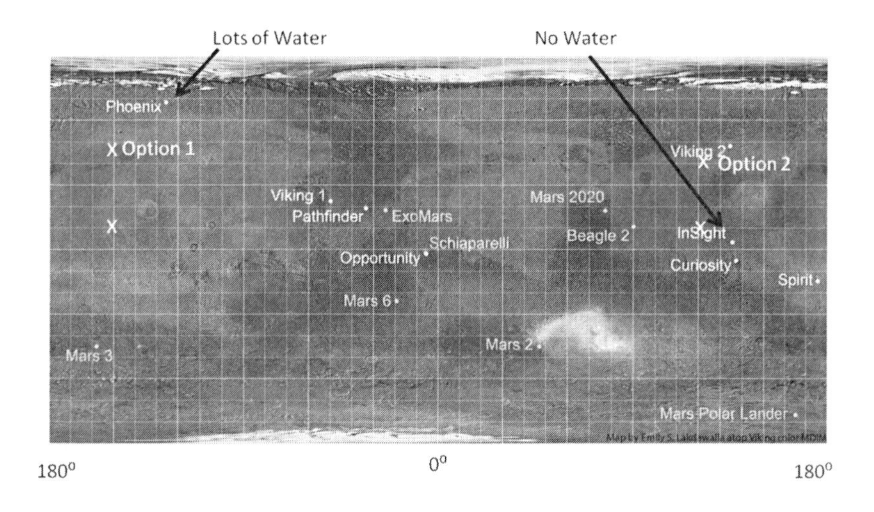

Figure 30: Past landing sites of Mars missions with options overlaid
Credits: JPL

This shows that near the option 2 southern point, it was very dry down to 300 meters. That is not what we hoped for, even though we are not assuming water near the equator. However, it is what we hope for eventually. In the case of option 1 the only data we have from actual landers is there was a lot of water at a slightly higher latitude. I have no idea why NASA did so much exploration south of the equator. That is not where we ever want to send people. It is higher elevations consistently, which means more radiation, less atmosphere to brake, and lower atmospheric pressure for fuel production. Perhaps their thought was just a sampling of the whole planet, but it wouldn't have been my thought if I were thinking of sending humans there later.

Having spent most of my career involved with NASA, I believe the issue is that they are very compartmentalized. The true scientists get to decide where the rovers go in their search for scientific answers and the history of Mars. Meanwhile, the manned mission folks are starving for specific water and landing site data, not digging into the history of Mars. That goes back a bit to the earlier discussion of knowledge for knowledge's sake versus how do I solve my problem.

Note that while Phoenix provided a lot of useful data, it died during the winter. The thought is that it may have frozen permanently in the harsh winter temperatures near the Arctic Circle.

The final figure we want to look at is the gamma radiation. How did we do on that? Are these locations that have relatively low radiation? After all, the astronauts or colonists are going to have to spend many hours a day outside or in a greenhouse, even if we provide shelter.

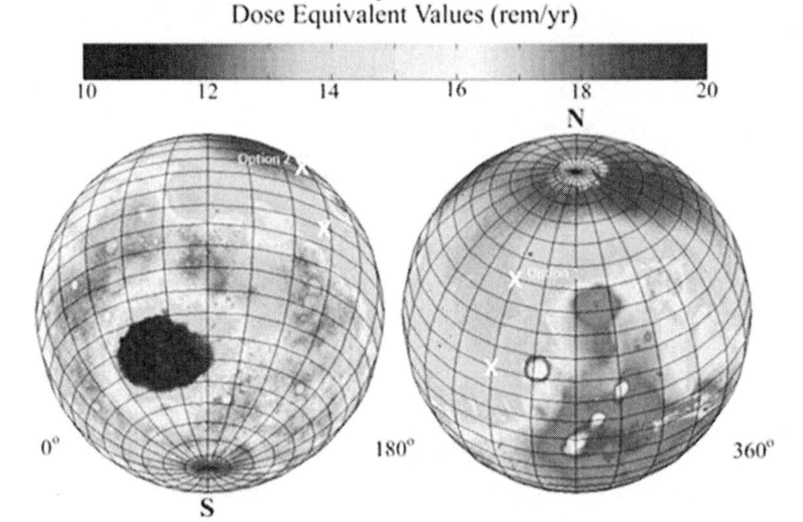

Figure 31: Landing locations vs. radiation levels
Credits: NASA

As we said back in chapter 5, we want to be in a medium grey area in the northern hemisphere. Figure 31 shows that we meet that criteria. The radiation levels are about as low as we can get without going into a deep canyon or crater that we may not be able to get out of.

One other significant thing to note is that the Option 1 southerly location is about 15 degrees west of Olympus Mons, the largest mountain in the solar system. It is the large white circle surrounded by a ring indicating high altitude. One article looking for geothermal power on Mars based on features similar to where geothermal power exists close to the surface on Earth pointed out the area around Olympus Mons as interesting and a possible area to find a geothermal source close to the surface [16]. This was based on features like lava flows and fissures on the surface around the mountain. It is also the largest volcano in the solar system so there was once a lot of lava there. Perhaps there still is underground.

The Trail

We can create a trail from our 45-degree location to our 15 degree location in order to deliver water. A bad solution is just having a vehicle with 15 tons of batteries that can make the trip. A better solution is to build charging stations along the way. How do we do this?

For starters, let's look at the distance. The diameter of Mars is 6772 km (average of polar and equatorial for a longitudinal path). The distance on a sphere is:

Distance = π * D * (degrees / 360)

So for 30 degrees, this is 1773 km (1100 miles). A 2020 Tesla Model S gets about 480 kilometers (298 miles) on a charge with a 75 kWh battery. It has a curb weight of 2215 kg (4483lbs.) and we can add in 4 passengers which is another 300 kg (662 lbs.). The 75 kWh battery weighs 480 kg (1,058 lbs.). Let's use this ratio to calculate range. (The newer batteries I believe have cooling and weigh more. Convective cooling doesn't work if you don't have a dense atmosphere and we may not need it at minus double digit temperatures anyway).

We will have a lightweight moon buggy-like vehicle to get over any rough spots, though it will be automated and have solar panels rather than any seats. It will probably weigh 500 kg (1100 lbs.) without battery, because it will be a tougher version intended to pull two trailers with water. (The original moon buggy was 260 kg including batteries, so this is at least triple). It is going to pull two trailers, each with 5 days of water or 3,575 kg (7,883 lbs.) of water. This is the same size as our fuel production water tanks, so we can fill one up from the other. Let's assume each trailer is 250 kg (551 lbs.) plus 105 kg (232 lbs.) for a water tank at 1 bar with a safety factor of 40 to handle the bumps, on a shock mount.

With one battery for the buggy, the total mass is about 8,840 kg (19,492 lbs.). On Mars, this is the equivalent of 3,359 kg (7,407 lbs.) on Earth due to the 38% gravity, at least as far as friction and dragging weight. This is about 33% more than the Tesla. The added weight probably makes up for wind resistance on Earth and traveling speed. We intend to travel about 20–25 kph maximum and don't have any significant wind resistance.

However, our terrain is not a smooth road. It is somewhere between packed soil and rocky. Let's say we get somewhere between one-half and one-third of the range. That is a guess. SpaceX can test it and adjust, but this is pretty conservative. That puts our range at 160–240 km (99–149 miles). Let's add some margin and say we will have a charging station every 148 km (92 miles) in case some sections are rougher than others. That means we need 11 charging stations to make the trip on 1 battery.

As far as these stations, humans may go back and forth and use them also. It is very credible that humans would service them. It is also very credible that humans might go between the locations for other reasons, such as a 45 degree worker moving to the 'good life' at 15 degrees or a 15 degree colonizer working a stint at the water station. Transportation capability between bases is a very good thing. Therefore, let's look at what humans might need 'on the road'.

Something we have talked about is dust storms. However, we know when they occur, near the start of summer in the southern hemisphere. We also can see them starting from Earth. Our astronauts should have days of warning. Also, heading out in a rover for days during this time to travel between locations is not something the astronauts would do, any more than someone would try to cross the south Atlantic in a small boat at the peak of hurricane season.

Another consideration is solar flares. However, the depots are too far apart to help with that. A solar flare starts with about 15 to 30 minutes warning on Mars (if you are looking at the Sun with a telescope) and lasts about 2 to 3 hours on average. Any protection from that has to be in the rover. It may be as simple as get out and crawl under the vehicle or the vehicle roof might even protect them. Adding layers of polyethylene and aluminum is pretty easy and quite effective. However, a solar flare is not going to kill anyone on Mars in a rover, but it is an elevated dose and to be avoided if possible.

Solar flares are often associated with a sun spot that forms. If a sunspot is seen from Earth, they should warn Mars to stay close to their shelter or at least close to their vehicle for a few days. An exceptionally severe event occurred in January 2005. A large spot formed and then it exploded 5 times over the next 5 days. The fifth one on January 20 was the most powerful since 1989. Its output was about 400 rem in Space at Earth. However, just the hull of an

Apollo capsule would have attenuated that to about 35 rem. That is not a dose you want but it would not kill you. About 300 rem is required to kill you[17].

At Mars, that dose in space would have been reduced to about 170 rem due to the distance. Density of radiation drops off as distance squared. Mars' atmosphere will attenuate it by another factor of at least 2 or 3 and just the solar panels on the roof would probably be more protection than the Apollo hull, so you are down in the range of perhaps 6 rem, which is the equivalent of 6 CAT scans. Crawling under the rover's batteries would reduce it even further. Therefore we do not need to worry about solar flares when it comes to the charging stations, but the rover designers may want to consider this.

Another question is do the humans just sleep in their rover? Perhaps the answer is yes. A manageable solution is a comfortable reclining seat and the closed rover provides a shirt-sleeve environment.

A final question is do we leave some supplies along the route just in case. I think the answer is yes, in case someone is stranded. We have plenty of payload capacity and a 30 day supply of food, water, and air for two people is only 384 kg (847 lbs.), including tankage. We could always go back and get it if we were desperate. This is all summarized in Table 17.

Charging Depot Requirements			
Item	**Mass(kg)**	**Quan**	**Total(kg)**
Solar Panels	3	100	300
Batteries	480	1	960
Charging System	100	1	100
Robotic Mate/Demat	100	1	100
Structure	300	1	300
Consumables	192	2	384
	Total		**2144**

Table 17: Charging depot mass requirements (per depot)

The water transport vehicle will be unmanned. It will be automated and move between 45 degrees and 15 degrees over about 6 days assuming 20 kph (12 mph) with a couple of hours charging at each depot. It will also have its own solar panels, about 4 m^2 per trailer and about4.6 m^2 on the vehicle. These would be used to provide a little bit of heat to keep the water from freezing, but they also provide security. If the vehicle were to experience unusually rough terrain or for some other reason not make it to the charging station, it could go about 20–35 km (12–22 miles) per day on its own to make it there without the need for rescue.

Once this set of charging stations are in place, we have a way to deliver water. We can land at 15 degrees without risk of being stranded.

This set of charging stations would be put in place on the first or second mission to 45 degrees and before anyone lands at 15 degrees. The intent would be to not only have this capability in place but also to pre-deliver some water to the 15 degree location so they have water on day 1. This plan assumes water rich at 45 degrees based on NASA data, which I believe is a given, but we are not certain until we get there.

How Does This Impact Us

The first thing that is obvious is we aren't going where we want to be in the first mission or two. That is true but done for the safety of the crew. Water may be very hard to come by near the equator. The other issue is solar variation.

Solar Variation

Mars is in an elliptical orbit, much more so than the Earth. It is much further from the Sun at it aphelion or farthest out (249.3 million kilometers) than at its perihelion or closest approach (206.7 million kilometers). This makes a significant difference in solar flux (power) as is shown in figure 32.

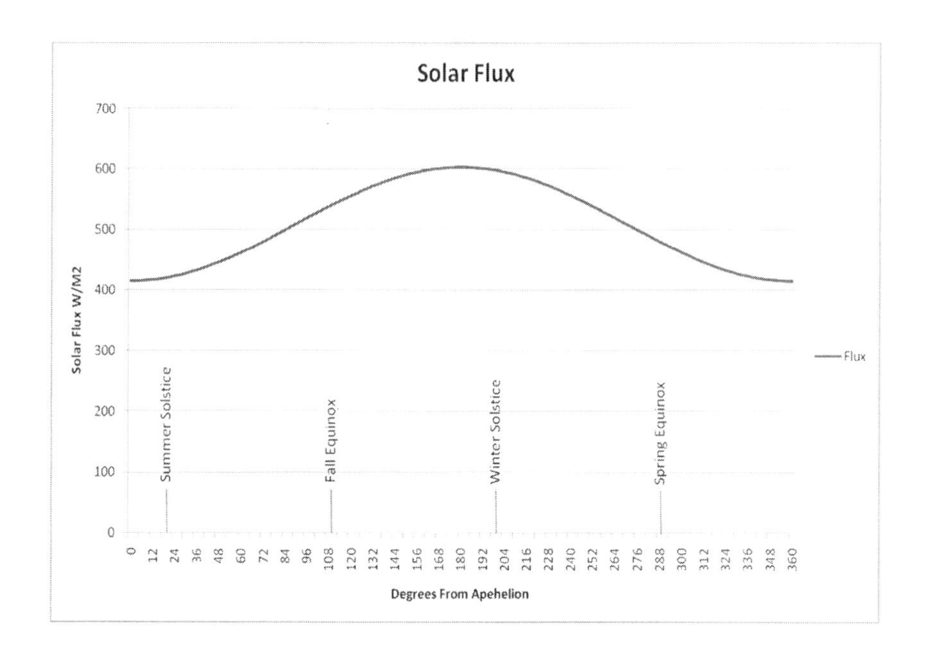

Figure 32: Solar flux by angle from Aphelion (farthest distance)

The good news is that the seasonal length of day tends to offset this since the low flux is in the spring and summer. The longer day compensation is much more significant at higher latitudes.

The 45-degree missions are the same latitude as Minneapolis. Their day-night cycle is far more seasonal. They have 16 hours of sunlight at the peak of summer (summer solstice) and only 8.7 hours of daylight at the peak of winter (winter solstice). Their daylight hour swing is shown in figure 33 (on next page).

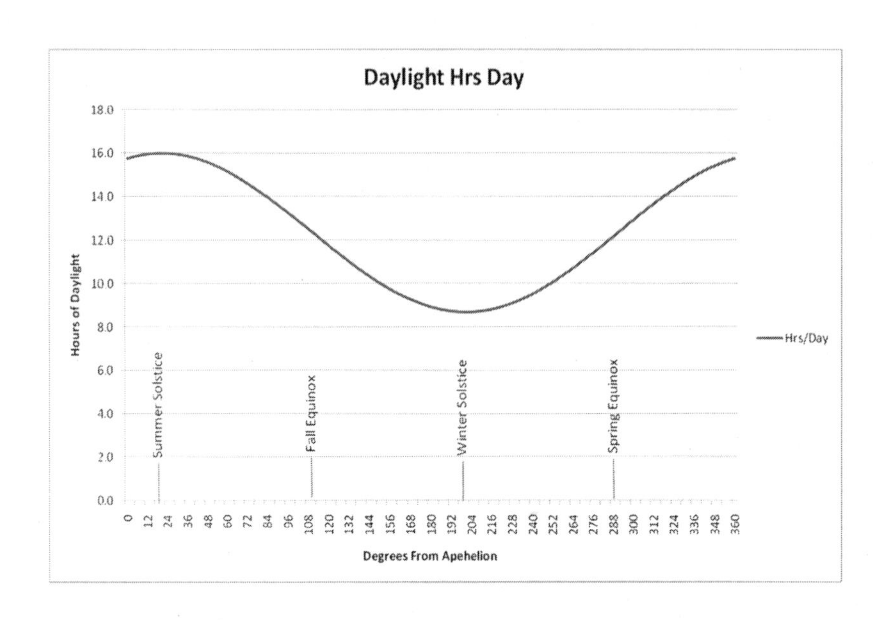

Figure 33: Daylight hours at 45 degrees latitude north on Mars

Notice the two sine waves are almost directly out of sync (off by 21 degrees from out of sync). This causes some cancelling out as Figure 34 shows. This plot is the actual power our solar panels will produce based on solar flux and length of daylight.

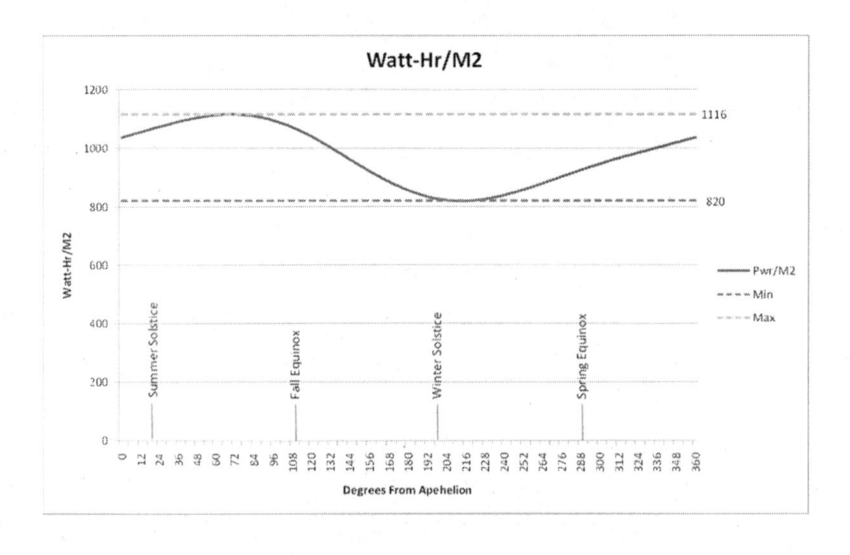

Figure 34: Actual watts per square meter produced per Mars day at 45 degrees north latitude

The power output is still highest in the summer due to the much longer days, but Mars is closer to the Sun in the winter. However, you can't just take a numerical average across this plot. A planet in an elliptical orbit moves slower when it is farther away from the sun. The orbit doesn't cover equal angles; it covers equal area as Keplar told us, as in the area of the pie wedge if you can think of slicing the ellipse like a pie. It moves through the angles more slowly in the spring and summer near the aphelion or furthest point out. On Mars Spring is 194 days, summer is 178 days, fall is 143 days, and winter is 154 days (all in Mars days). The net result will still give us 948 W-Hr/m2 per Earth day or 974 W-Hr/m2 per Martian day just like we discussed in chapter 6. It just varies by season. Figure 34 is per Martian day.

Our length of stay is supposed to be 550 days. That is 0.8006 Mars years, almost exactly 80% of a year.

Arriving at Mars somewhere around the fall equinox would not be ideal. We would miss the summer peak as our missing 20%. Arriving just before the spring equinox would be ideal. We would miss the winter. However, we don't get to choose this. The launch window happens every 2 years 49 days so we get one shot every couple of years.

The difference in power output is not a show stopper. It means we will have nominal power on average over a mission plus or minus 3.4%. If the ship is arriving at just the wrong time you take 3.4% more solar panels and produce the same amount of power as planned or just finish fuel production 16 days late, but we have 70 days margin. That is another 367 m2 or 1.1 tons for the absolute worst possible timing. For the best case you either leave a few panels home or finish fuel production 16 days early, probably either the latter or use the extra power for something else. You might need the panels for the next mission.

The seasonal variation in power output does, of course, drive being able to throttle the fuel production processes. We must have adjustable input to the Sabatier reactor and have thermostat driven condensers because they will run at less than full capacity in the winter.

The computer control system will have to adjust power usage to power availability. This means running electrolysis and water production for fewer hours in the winter and more in the summer. That in itself has no impact on electrolysis or water production. They just produce more fuel in the summer than the winter because they have more daylight. The issue comes with the

Sabatier trying to keep up. It is running 24/7 but the power ramps up and down about 14.8% from the average. However, this can be accounted for. In the summer we ramp the Sabatier process up about 13.3% and ramp the electrolysis process only up about 13.3% instead of 14.8% from its average production rate per day. The Sabatier process is only about 11% of the power usage, even though it runs 24/7, so this split would balance the two processes in their output and they must be synchronized on hydrogen production. We have enough buffer for daily differences, but not seasonal.

Note that this power ramp up is only to take advantage of the additional power available due to longer days. The system would ramp down this amount in the winter from the average. The power availability and production would follow a sign wave. Over the span of a year it would average out but we are not planning to stay a whole year so when we arrive matters.

This same principle applies at fifteen degrees, but there is not a large day length change to more than offset it. At 15 degrees north the daylight hours look like figure 35.

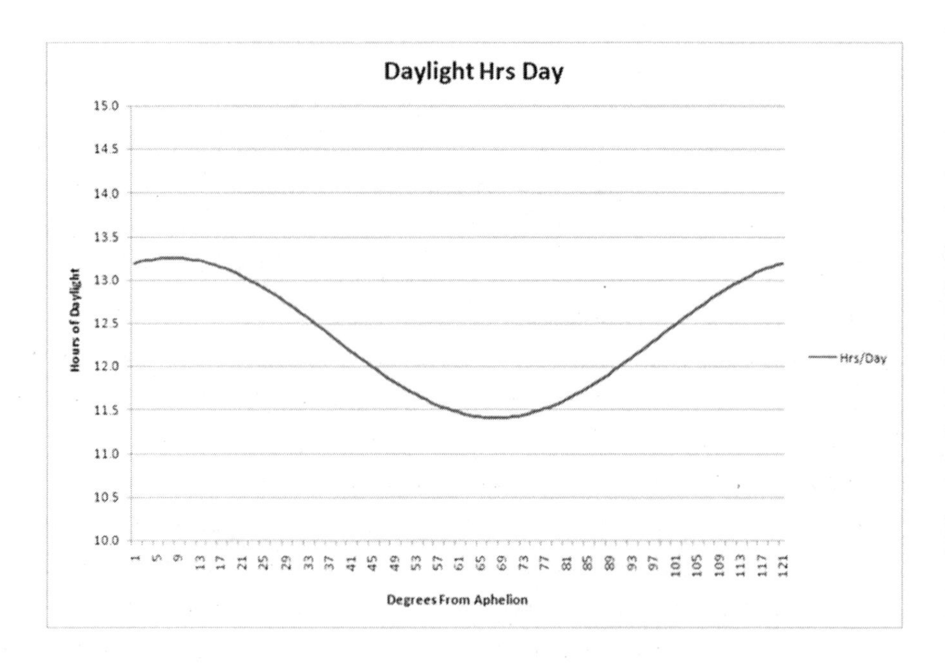

Figure 35: Daylight hours at 15 degrees latitude north on Mars

This is again in opposition to the solar flux, not enough to offset it but it does dampen it a little. The net power our solar panels will produce is shown in Figure 36.

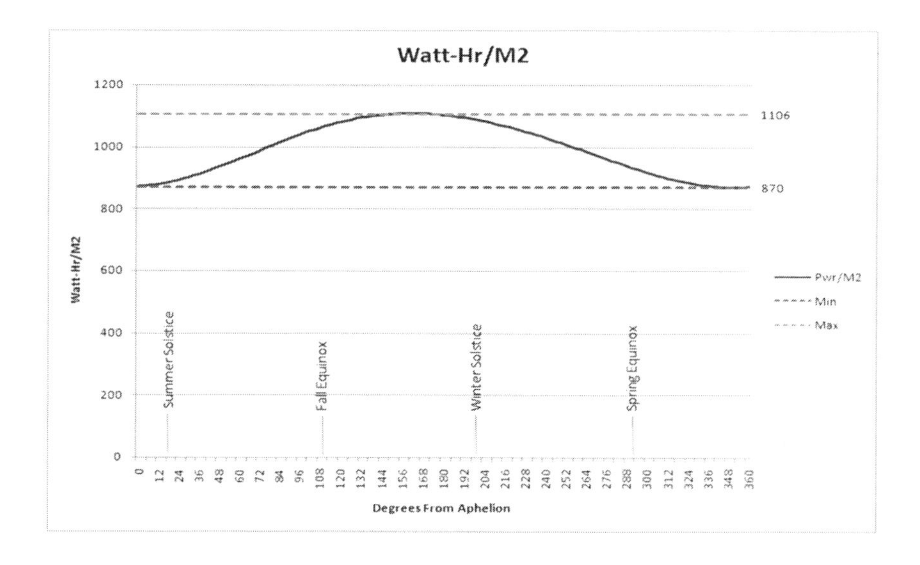

Figure 36: Actual watts per square meter produced per day at 15 degrees north latitude

It peaks in late fall at or near the perihelion (closest point to the sun).

This has the same implications for power usage but not with the same timing. At 15 degrees latitude you would prefer not to arrive near the winter solstice and would prefer to arrive in the early summer. The swing is about 3.5% and can be managed, just like the swing on the seasons at 45 degrees. Somewhere at a latitude in between is an almost flat seasonal distribution but not zero, since the two effects are not quite out of sync. Seasons, arrival dates, and benefit or loss in solar power is shown in Table 18.

The launch and arrival dates were provided using the Ames Research Center transfer orbit database. All are supposedly the minimum energy transfers for that year of about 180 days. (They actually seem to like 176-day transfers, but this is close enough for our purpose.) Launch windows are a little over two years apart as discussed back in chapter 4. This table is intended to show the effect of latitude and season. Actual plus or minus was estimated. Plus means you have a slight excess of power. Minus means having slightly less than our previous calculation of average power. This is due to the seasonal variation just discussed.

Spring Equinox	**Summer Solstice**	**Fall Solstice**	**Winter Solstice**	**Launch Date**	**Arrival Date**	**Solar Power 45 Degrees**	**Solar Power 15 Degrees**	**Global Storm Likely**
5/10/2017	11/20/2017	5/17/2018	10/7/2018					Happened
3/28/2019	10/8/2019	4/3/2020	8/24/2020					
2/12/2021	8/25/2021	2/19/2022	7/12/2022					
12/31/2022	7/13/2023	1/7/2024	5/29/2024					Yes
11/17/2024	5/30/2025	11/24/2025	4/16/2026					
10/5/2026	4/17/2027	10/12/2027	3/3/2028	12/10/2026	6/4/2027	Minus 1%		
8/22/2028	3/4/2029	8/29/2029	1/19/2030	1/16/2029	7/11/2029	Minus 3%		Yes
7/10/2030	1/20/2031	7/17/2031	12/7/2031	2/23/2031	8/18/2031	Minus 2%	Break even	
5/27/2032	12/7/2032	6/3/2033	10/24/2033	4/17/2023	10/10/2033	Plus 1-2%	Minus 3%	
4/14/2034	10/25/2034	4/21/2035	9/11/2035					Yes
3/1/2036	9/11/2036	3/8/2037	7/29/2037	7/12/2035	1/4/2036	Plus 3%	Minus 1%	
1/17/2038	7/30/2038	1/24/2039	6/16/2039	9/3/2037	2/26/2038	Plus 2%	Plus 1%	
12/5/2039	6/16/2040	12/11/2040	5/3/2041	10/27/2039	4/4/2040	Plus 1%	Plus 2-3%	Yes
10/22/2041	5/4/2042	10/29/2042	3/21/2043	11/28/2041	5/23/2042	Break even	Plus 3.5%	

Table 18: Mars seasons (Earth dates) and mission arrival dates with power production impacts

This shows the impact per mission as far as solar flux average over the roughly 550-day stay on Mars. The cycle repeats about every 7.7 Mars years. This is because the launch window occurs every 1.13 Mars years, so the arrival date is a little later each year on a Mars calendar but eventually loops around after missing one year entirely. Notice that the 2027 arrival has nearly the same impact as the 2042 arrival, which is 8 Mars years later.

Something of interest also might be the potential impact of dust storms. They typically start somewhere near winter solstice in the northern hemisphere and could impact fuel production for days or weeks. The global storms usually, but not always happen every third year. Sometimes this event just doesn't happen. It occurred in 2007 and 2018, but was expected to occur once in between and didn't, at its 5.5 Earth year interval.

In the southern hemisphere, the effects of daylight hours and perihelion would actually compound, which is why we get the extreme boil off at the South Pole. It has its summer at perihelion, or closest approach to the sun. It also means the seasonal variation would be much larger in the southern hemisphere.

Now that we have talked about where we will go and the implications on solar power and other things, let's discuss the things we need once on Mars.

The Infrastructure for Colonization

The purpose of going to Mars is for humans to first begin to occupy, permanently, another planet in the solar system. The astronauts or pilgrims, whatever you might call them, are going to be very historically unique human beings.

Buzz Aldrin, Astronaut

Habitats

For early explorers, the habitat may be the ship they arrived on. It is certainly adequate for a 550-day stay since it worked for the 180-day trip in and they would bring at least 2-1/2 years' worth of supplies. It also has a complete recycling system for oxygen and water. However, as we start to leave colonists on Mars, they will need habitats on the planet.

Most authors want to put habitats underground. The reason is that Mars has very little atmosphere to shield explorers and colonists from cosmic rays and solar radiation. In order to shield gamma radiation to what are considered safe levels takes some shielding. Part of the issue with safe levels is who decides what is safe. Astronauts in the Space Station receive more radiation than is allowed for even nuclear power workers. The nuclear power worker level is a very conservative 20 mSv (2 rem) per year from all forms of radiation.

Cosmic and solar rays are reduced by Earth's atmosphere. The average dose on Earth is about 2.4 mSv (0.240 rem) per year. We can never get the dose to zero. In fact we wouldn't want to. Radiation is believed to be involved with mutation, which was the driving force for the development of all the varied life we have on Earth and will continue to evolve us to adapt to our environment.

This is the average dose. However, some locations have much higher radiation. Ramsar, Iran for example is one of many places with higher radiation. It has 30–50 times the normal level. This level is caused by the use of naturally radioactive limestone for construction and the presence of radon. However, the people there do not have elevated rates of cancer, which is the primary negative outcome of slow, steady radiation doses[18][19]. This implies that humans can tolerate a much higher level. In fact, just going to Mars with a 2.5 year round trip will expose astronauts to pretty high levels, but actually less than some Space Station astronauts have experienced already. Many sources now are suggesting that 100 mSv (about the same as Ramsar) probably has very little effect on humans[20]. In order to shield long-term residents to this level requires about 1 to 1.5 meters of soil[20][21]. Hence the 'Let's go underground' sentiment.

Looking at the map of Mars in Figure7 back in Chapter 5, it shows that the surface of Mars has a range of 10–20 rem/year (100–200 mSv). In that chapter, we said we wanted to settle in a blue area not a red one, so that would put us in the 100–140 mSv (10–14 rem) range already, so we are close.

First of all, that says it is safe to work in a greenhouse during the day. However, to bring our total dose down to below the level of 100 mSv, let's cover our habitat in soil.

References say the radiation is already attenuated in the first 15 cm by 70% [22]. However, the issue with that is that the collisions of gamma rays with the material cause a scattering of other particles. They recommend 2,000 g/cm2 of material to capture these other particles. That would mean 1.32 m (4.4 ft) of regolith using the density of dry Earth soil. I believe that is where the 1 to 1.5 m comes from.

It is possible to make concrete on Mars and obviously we will when there is a need for it and we locate the materials. However, I think for the near term, habitats will be inflatable. Otherwise, the air we need to breathe would seep out through the soil (or blow it away, as in move the dirt if it were not porous).

Option 1

There are only 2 shapes to really consider. The most efficient shape for any inflated object is a sphere in terms of surface area to volume. In fact, if you blow up anything, the walls will bulge out and try to form a sphere, so let's consider a sphere. It is the least surface area per volume. Whether we have it completely buried or only partially, we will probably fill it up halfway with regolith so we have a level floor to walk on. Therefore, it will look more or less like Figure37.

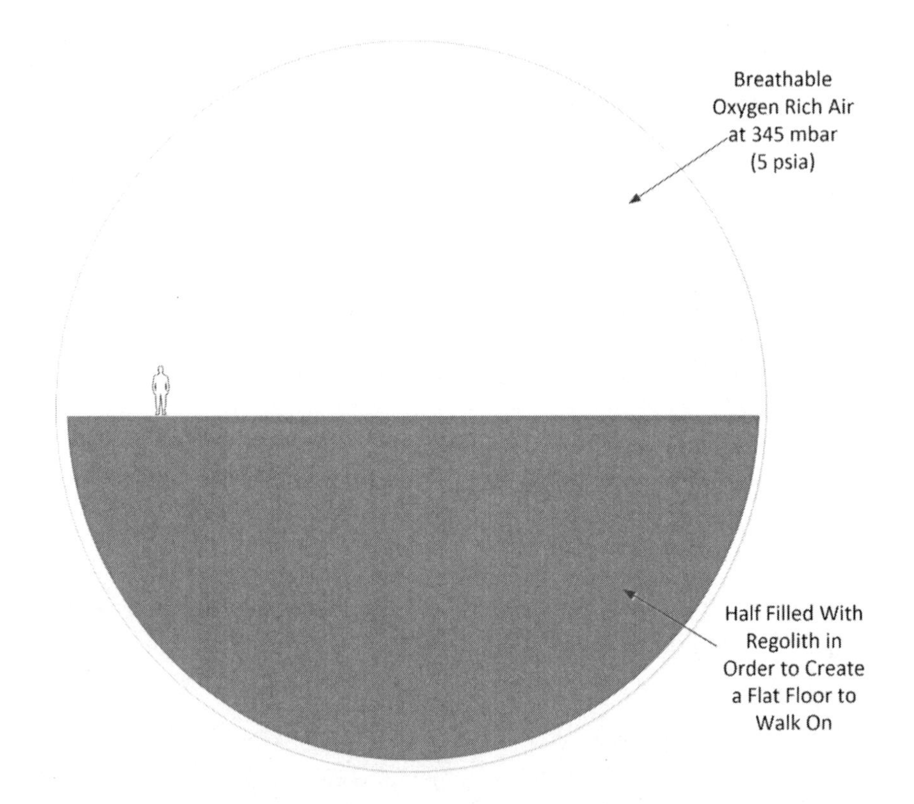

Figure 37: An inflatable habitat could be used on Mars

The 'balloon' would probably be made of either Kevlar (not transparent) or of a strong clear plastic, like clear polyethylene with a Kevlar or carbon fiber webbing to reinforce it. The reason for the reinforcement is that while polyethylene is very strong for momentary impacts, it tends to stretch if held under heavy loads for long periods. The Kevlar or carbon fibers could be used to take most of the load and prevent the slow creep (permanent stretching).

We could bury it completely and have some kind of passageway out. However, if we were to only bury the sphere halfway under the ground we would be able to exit at ground level. We would still cover it in regolith for shielding from gamma rays, but provide an exit tunnel in order to leave the habitat. This tunnel might lead to an air lock and the Martian surface or it might just lead to a greenhouse. The half-buried version might look like Figure 38 below.

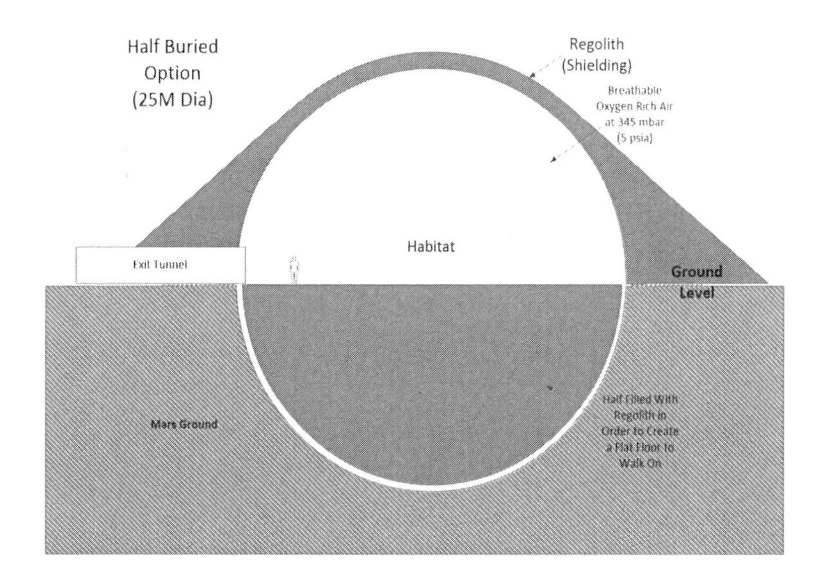

Figure 38: Rather than completely burying a habitat, it could be only half-buried and exit to ground level

Another variation to this theme is to provide roof supports all the way around the structure and leave the entire outside lower edge open, providing a view of the planet. I would prefer that option if given a choice (as a resident) but the roof support structure is likely to be quite a bit heavier than tunnel(s). Therefore, let's assume the half-buried option above for calculating mass requirements.

If the habitat is pressurized, the next question is how much and with what. The thickness of the shell is determined by the pressure. It has to be able to withstand the forces generated by the gas pressure. The pressure at sea level on earth is 1013 mbar (14.7 psia). However, it is mostly inert nitrogen. Let's remove some of that. Humans don't really need sea level pressure; they just need a significant partial pressure of oxygen. A wiser choice would be perhaps 5 psia or 345 mbar that is 62% oxygen. This is the same amount of oxygen, just a lot less nitrogen. We will discuss this more when we get to greenhouses. For our discussion here we just need a pressure to work with to calculate mass in our structure and also its ability to support weight.

The atmospheric level of 345 mbar (5 psia) is adequate to support about 6M (20 ft) of regolith in the light Mars gravity. The actual regolith depth would be about 1 to 1.5 M (3–5 ft) on the top, where it is being directly supported by the habitat. On the sides much of the weight is carried by the regolith below it, rather than by the side which is approaching vertical as we go downward. The tunnel is an exception. Based on the slope I drew in figure 38, the regolith depth on the tunnel is up to 5.5M. This slope is more a function of gravity and soil characteristics, rather than something we have direct control over so I am going to assume we need some reinforcement in the first meter or two of the tunnel. I am going to make an allowance for it but not actually design it.

In this version, the shell is assumed to be a Kevlar balloon since we don't need transparency. Kevlar would only have to be about one-eighth of a mm thick for a 25M diameter structure (5 mils) with a safety factor (SF) of 4 at 345 mbar. This brings the shell mass in at about a quarter of a ton. However, for the human habitat I am going to assume we double this thickness to assure no punctures due to sharp pieces in the regolith, which could create point loads higher than the 345 mbar air pressure. This will also be assumed for the bottom on greenhouse structures (SF = 8). It also probably has to have a thin plastic (polyethylene?) liner.

Kevlar is normally a woven material. It probably won't make an airtight structure, so the liner would provide a seal while the Kevlar would provide the strength. Kevlar is a very tough material used in bullet proof vests so it should be able to hold up even buried in regolith, though testing could prove this out. We also might use a spray foam on the inside of the bottom half of the plastic layer just to provide something crushable to protect the polyethylene balloon from any sharp edges in the regolith.

Assuming a sphere is chosen, it may be possible to deviate slightly from a sphere, as in a shallower half on the portion buried. This was considered, but anything like just going to a straight wall down a couple of meters and burying that didn't work. The edges would pull up under the low Mars gravity and tend to pile the dirt in the middle as the sides lift up. However, it is believed that there is a possible solution in between where the weight of the regolith offsets the tendency for the 'balloon' to want to go spherical and a slightly shortened hemisphere on the bottom and/or top is possible. All calculations that follow do not include this, they assume a full sphere. This is conservative and actual outcomes could be slightly better.

The sphere will be buried halfway in the ground and then filled halfway with dirt so we are farming or walking around on the largest possible area at ground level. In living areas we will want to seal over the dirt with plastic as a water barrier and then add flooring, probably something like vinyl with a carbon fiber or aluminum backing, which could make for a firm and somewhat sound absorbent floor but also be lightweight. The interior structure will probably be modular and lightweight, as in plastic and perhaps some carbon fiber or metals where strength is needed, including to support a second floor. The thought is that the habitat will house about 14 people with a diameter of 25 meters. This means it is 12.5 meters (41 ft) high in the middle. That is room for more than 2 floors over the middle portion if desired.

This area within the habitat dome will not just be personal space. It will include some common areas, perhaps even a common kitchen, a living room type area, laundry, and other needed facilities. It will also include space for storing food, several tons of it. Farmers collect their crops at certain times of the year and then have to eat them for many months. I remember both of my grandmothers who lived on farms doing a lot of canning in the fall. Some crops like potatoes can be stored for months in a cool dark place, but we have to make it last until the next crop, so we may do many things to preserve food. This could include freezing, canning, and probably some other options, depending on the vegetable. Most fruits can be frozen and most vegetables do quite well with canning and can last months or even years once canned, but all of this takes space, whether freezer space or shelf space.

If we are trying to get people to move to Mars, there are things we want to include. That means private apartments for each couple or family. It means sound-insulated walls that provide some privacy. It means a private bath or at least a toilet and sink. It means access to open areas (like the greenhouses). It means some kind of kitchen facilities and laundry, even if it is shared. It has to be at least as nice as Space Station facilities or (a lot) better.

The floor space on the ground level is 491 m^2 (5,285 ft^2). That is quite a bit, but we are going to have 14 people and their food for months in it. It would probably be worthwhile to add some structure and put living quarters / apartments on the second floor, possibly along with some common areas. The first floor then could be used mostly for food processing and storage, as well as storage for things like EVA suits, and a place for equipment like sewage treatment, water tanks, and other things needed for 100% recycling.

Note that this structure cannot touch the walls or it could possibly puncture them if shifting occurs.

Having a second floor means building walls sturdy enough to support it and beams across the floor of the second floor. However, materials we are flying to Mars will be as lightweight as possible. Mass is more precious than materials like honeycomb aluminum floors or carbon fiber beams. We will also want some sound insulation, which can be very lightweight and cheap, like Styrofoam inside walls, for example.

These additions would make this a nice and very functionally adequate space for the new colonists.

Option 2

The other shape of interest is long cylinders. While they have more surface area per volume, the stresses are lower so walls can be thinner and the overall structure is just as light. Cylinders would also be easier to assemble, since you aren't trying to get regolith 12.5 meters (41 ft) up in the air and the mass over a given area is less. It also does not drive reinforcing its tunnels.

As Figure 39 shows, the concept is basically the same. It is just a long cylinder with a smaller diameter, probably about 8 meters (26 ft).

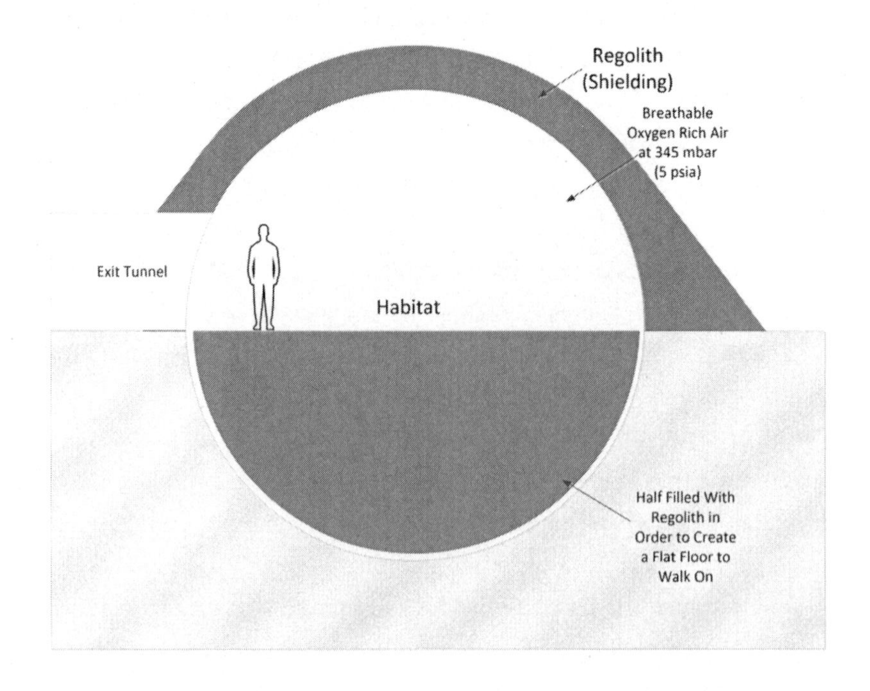

Figure 39: Cross-section on a long cylinder habitat

This option is actually quite viable. No second floor is practical, but at 8 meters diameter and 90 meters (295 ft) long it offers more floor space than the 25 meter sphere, making a second floor space unnecessary. These options will be revisited when we get to greenhouses.

Being a colonist will mean working to sustain yourself. It is probable that everyone will have jobs. Perhaps 2 or 3 people are the farmers, who tend the crops. Perhaps 1 or 2 take on the role of cooking and canning. Someone else might explore for new water sources by performing seismic studies in different areas. There will be plenty to do.

While we do have a lot of technology that early explorers in America lacked, colonists will be mostly on their own and providing everything they need to live on. In some ways, it will be like pioneer days in America, where people worked from sunrise to sunset trying to eke out a living. I am not sure how many people find that desirable, but there certainly are some who seek adventure and will accept what it takes. People colonized in America for many reasons; some to get away from what they left behind; some seeking adventure; some just wanting the freedom of being in a new place with no one telling you what to do. People will go to Mars for some of the same reasons.

Greenhouses

People need to grow food to stay alive. We are not going to send food from Earth forever. This requires greenhouses on Mars. People on Mars will be vegans, whether they want to or not. Raising animals requires way too much acreage, as in 10 or 20 times the area for the same calories that you can get from vegetables. Growing vegetables will be done in greenhouses with elevated CO_2, but otherwise a (Mars) nominal 5 psia atmosphere with 62% oxygen so that the humans can walk around without spacesuits, as shown in Table 19. The higher level of CO_2 enhances plant growth [23].

Gas	Percent	Units	Units
Oxygen	62.0%	214 mbar (3.1 psi)	Same partial pressure as sea level Earth
CO_2	0.1%	1,000 PPM	This is over 2 times Earth level but great for plants
Water	4.9%	17 mbar (0.25 psi)	75% humidity @ 20 deg C
N_2	20.7%	71 mbar (1.04 psi)	
Argon	12.3%	42.4 mbar (0.62 psi)	
Total	100.0%	345 mbar (5 psi)	
(Mostly just removed a lot of N_2 but argon higher)		(Note that mix is same oxygen content but elevated CO_2 for plants within a safe level for humans)	

Table 19: Standard atmosphere on Mars in habitats and greenhouses

This, or something close to it, will be the normal atmosphere for every habitat and greenhouse on Mars. It keeps the same amount of oxygen but eliminates most of the (inert) nitrogen. This makes enclosures lighter while maintaining a shirt-sleeve environment. The reason for the balance in nitrogen and argon is that is what we vent from the Sabatier reaction after we condense out the methane. Why not capture some of it? It is a good natural source. The water, CO_2, and Methane are condensed out of the exhaust. The CO and O_2 in the atmosphere would react in the Sabatier[24].

This leaves us with a mixture remaining that is composed of about 63% nitrogen and 37% argon. If we can capture and use these it is a free source for our atmosphere. Recall that we are producing large amounts of O_2 and water. CO_2 is free by adding in a little bit of Mars atmosphere or from the condensed CO_2 and we don't want to bring along our 33% inert gases from Earth. It would be a couple of tons of gas plus the tanks per 14 person community.

It initially appeared feasible to go as high as 0.5% CO_2 based on studies with humans that said they are fine and plants thrive. However, a recent article in airqualitynews.com claims that other studies show anything over 1,000 PPM (0.1%) has adverse effects on humans that are not immediately obvious such as inflammation and kidney calcification. Therefore, let's limit it to 0.1%. Lots of CO_2 is great for the plants and encourages photosynthesis so we do want to go as high as is safe for humans, especially since we are already behind the curve on plant growth with just over half the sunlight we have on Earth. At least on Earth elevated levels of CO_2 like this increase plant growth by about 20%. Many commercial greenhouses pump in CO_2 for this benefit (and to offset the deficit the plants tend to create).

There are several general topics or concerns relative to green houses on Mars. Let's address them one by one.

Radiation

One of the first things to say is there was a research paper that said greenhouses on Mars are impossible because of gamma rays from cosmic radiation. That has been debunked. The researcher that reported this put radioactive material next to the plants. This was uncontrolled as far as the exposure and probably not the same level of gamma radiation as cosmic rays on Mars surface.

The magnetosphere blocks some ionized particles but has no effect on gamma rays. The space station is experiencing what Mars does or a little more due to Mars' atmosphere eliminating some of the radiation and being closer to the Sun. Plants grow very well on the space station with no issues over multiple generations of plants[25][26]. Therefore, gamma radiation is not an issue for the plants.

UV Rays

The UV rays on the surface of Mars are much higher than on Earth. They would be fatal to the plants and harmful to humans without protection. However, the solution is Silica Aerogel.

The reason for Silica Aerogel is that it is transparent, an excellent insulator, and also absorbs almost all of the harmful UV rays from the Sun. Otherwise the UV rays would kill the plants, since Mars' atmosphere blocks very little of it. The Silica Aerogel would make the environment safe for the plants and also much safer for humans. A study on the subject of Silica Aerogel for use on Mars found that 2–3 cm (1 in +/-0.2) on the greenhouse would not only filter out nearly all of the UV rays, but also get the temperature inside permanently above freezing, even though it is cold on Mars [27][28]. They had both theoretical and experimental evidence.

Silica Aerogel is an excellent filter specifically of UV while being almost optically transparent to visible light. The transmission spectrum is shown in Figure 40.

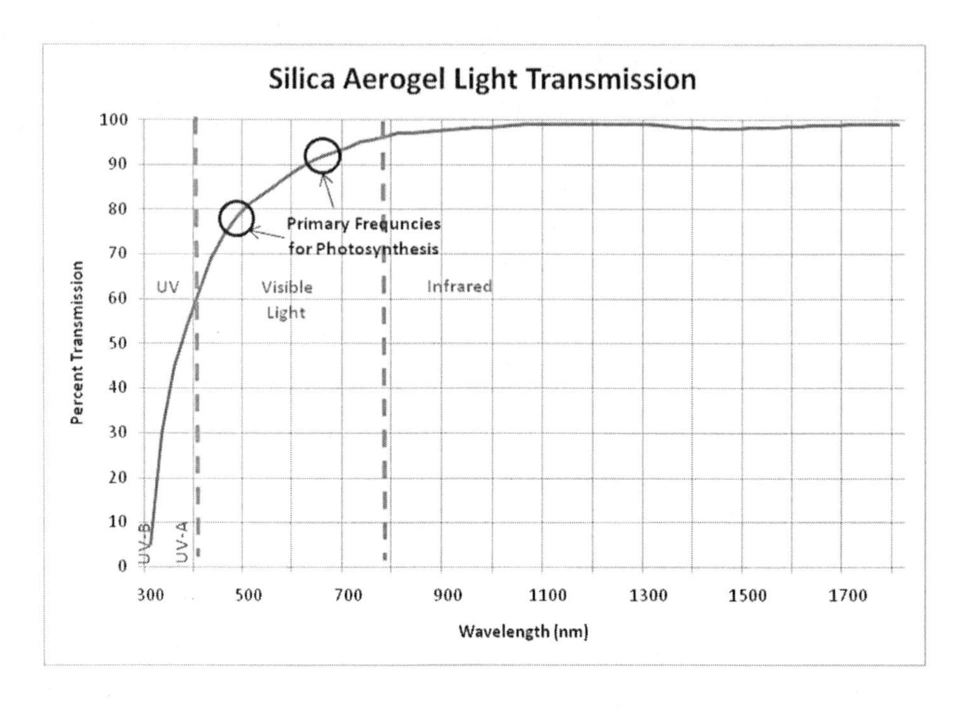

Figure 40: Transmission Spectrum of Silica Aerogel

Perchlorates

On Mars, the regolith (dirt) contains perchlorates. They are harmful to humans because they block iodine and prevent hormone releases from the thyroid. This is reversible for an adult who later leaves the environment, but probably not reversible for an infant due to lost development, and we will raise children on Mars. We have to remove this from the soil we grow our food in. A pretty good option is to release Earth bacteria in the greenhouse that consume the perchlorates. The only issue is they leave salts that are fatal to plants. The perchlorates and the toxic (to plants, not us) salts are very water soluble, so either way water is probably involved in removing the things we don't want in the surface soils.

That may not be a huge issue since we are going to water our plants anyway, but a little research is needed into how much water is required. Either way, we will have access to water, so this isn't a show stopper, just a bump along the way. As Matt Damon said in *The Martian* (paraphrased), "You have to engineer the hell out of it. If you solve enough problems, you get to come home." Or, in our case, you get to survive in a somewhat hostile environment, but one that has the resources we need.

Sunlight

The sunlight on Mars is roughly half of what it is on Earth, actually 52% as was discussed relative to solar power. This is an issue for some plants according to at least one study[29]. While several food plants did well, others like corn and tomatoes did not. One solution to the reduced sunlight is to add mirrors outside of the greenhouses to reflect light, which would add sunlight but also add heat and raise the inside temperature a few more degrees. These mirrors could be as simple as a very lightweight carbon-fiber frame with Mylar for the mirror. (Think space blanket light.)

They would be set up to focus light on the plants, especially those that require more sunlight. We might also have to select our crops carefully since this still would probably not get us up to the full level of Earth sunlight, although we could supplement with LED grow lights in very limited areas for specific crops we really want or need in our diet that don't grow well in perhaps 60–80% of Earth sunlight.

Food Density

The area needed for a greenhouse is about 364 M2 (4,000 ft2) per person according to a 1970's study [30]. They also want to say add the same for paths and storage, but we are using habitat area for storage and will have minimal paths. We have no heavy farm equipment that requires open rows. Our greenhouses are sized for 3.5 people each. We also may get more production due to no real winter, allowing multiple crops per year.

In northern climates, there is one growing season. However, in tropical climates, two or even three are possible. With the long Martian year (687 days), it might be possible to have several growing seasons, perhaps four or five. We will not have a colder season near the equator, or not noticeably so anyway. The sun intensity will vary some due to Mars' elliptical orbit but we believe we can keep the internal climate warm enough.

Density of food production is another thing to be considered. While all of the green vegetables are important for nutrition and will be included, there may be a focus on roots, like potatoes. They produce much more mass and calories per square meter than plants like peas or green beans. Therefore, we will produce a lot of potatoes and other similar roots for calories. Protein may also be an issue leading to production of legumes like black beans, soy beans, and lentils. Of course, we want tomatoes and corn and squash, but some of those at least may be minority crops, though corn is perhaps promising.

Many articles also suggest 100% plant usage yields the best results, like dandelions, which are 100% edible. I am not a PhD in agriculture, so I will leave it there. People with specific knowledge will make the best decisions, not just someone who has read a dozen articles on the topic. The point is though, that we can live on Mars and raise enough crops to live on.

The plants will be OK with the gamma rays, we can filter out the UV, we can wash the bad stuff away, and we will have enough to eat. We engineered the hell out of it, as Matt Damon would say, so let's look at the details.

Option 1: Spherical

A possible configuration for a greenhouse is shown in Figure 41. It is sized to feed 3.5 people, so each habitat will need four of them. It is 40 meters diameter, which provides 1256 m2 (0.31 acres) of farmland per greenhouse or 5,024 m2 (1.24 acres) total.

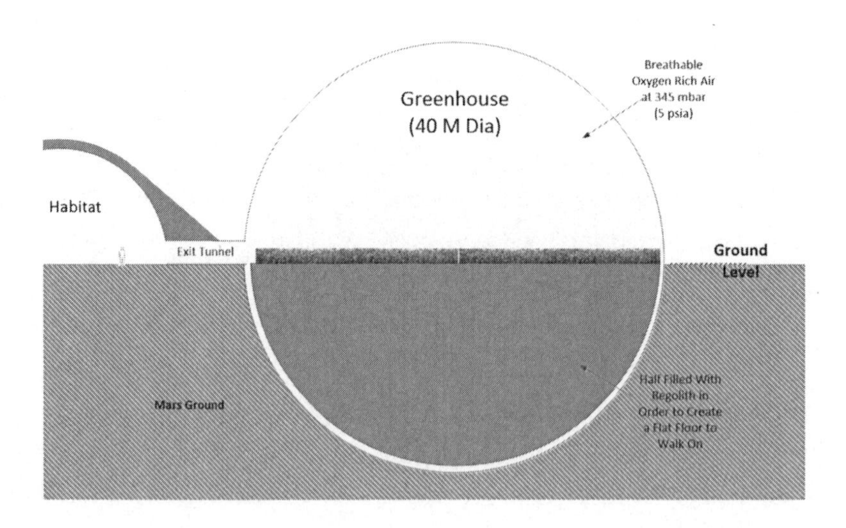

Figure 41: Greenhouse might be a 40m diameter sphere half-buried in order to have maximum usable area. It could be connected to the habitat by a tunnel.

The enclosure, which has to contain 5 psia, or about 3,582 kg force (7900 lbs.) per square meter and be transparent, would be best made from a strong, clear plastic like polyethylene with a Kevlar mesh on the outside woven like a fishnet in order to minimize light blockage. This could result in a durable lightweight shell. The shell is buried in the ground halfway and filled with dirt halfway to create the greenhouse. It also would probably include 2-3 cm of Silica Aerogel and an outer thin layer of a clear plastic on the exposed surfaces. The bottom, however, would be only Kevlar with a thin polyethylene liner, just like the habitat, since that is very strong and light and does not need to be transparent.

When put together with the spherical habitat as described previously, the facility for 14 people would look something like Figure 42.

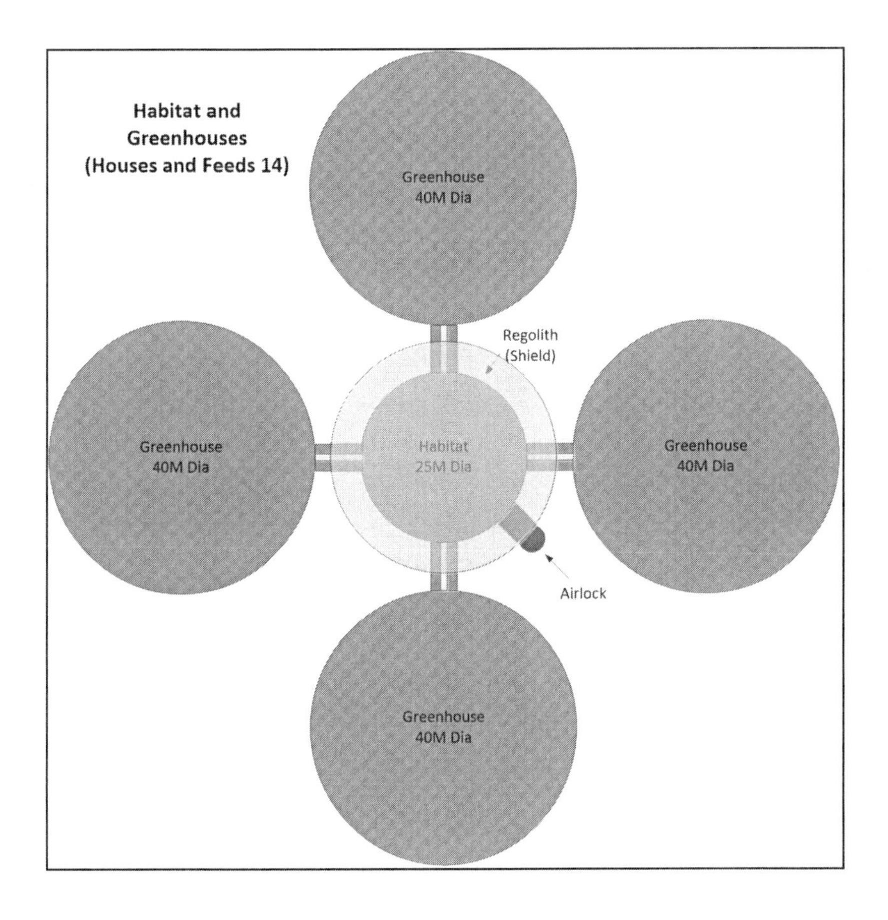

Figure 42: Option 1 – A possible home to feed and house 14 people with spherical domes

The idea here is open tunnels between the green houses and the habitat, though I did allow for a pressure door on each tunnel, just in case of a puncture.

However, a small puncture could be repaired with an interior patch. A small puncture would take days or even weeks to let enough air out to collapse a greenhouse even if isolated so there is plenty of time to react. The space station has similar repair kits in case of micro-meteorites.

The habitat also has a fifth tunnel leading to an airlock. This allows access to the outside.

Option 2: Cylindrical

If it is determined that 25–40 meter spheres are impractical because of the extreme depth and height, as in how do you dig 20 meters (66 ft) deep and how do you get regolith 12.5 meters (41 ft) up on top of the habitat, there are other options that were looked at. The best is a cylindrical option, where all structures are cylinders 8 to 10 meters in diameter (half of it buried) but quite long, as in approximately 100 meters or shorter and more of them. An example of this concept is shown in Figure 43. The greenhouses are 8.5 meters wide (diameter) by 107 meters long, including hemisphere end caps. The six of them are about the same farm area and sufficient to feed 14 people based on the same 4,000 ft2 (372 m2) per person. It is actually more square meters of material but does not have to be as thick as the 40 meter greenhouse material due to lower stresses, though I stuck with the safety factor of 8 on any surface touching regolith. The length of the buildings would be pointed east-west in order to benefit from reflectors and maximize the mid-day sun. In any other configuration the reflectors are only at the correct angle once per day.

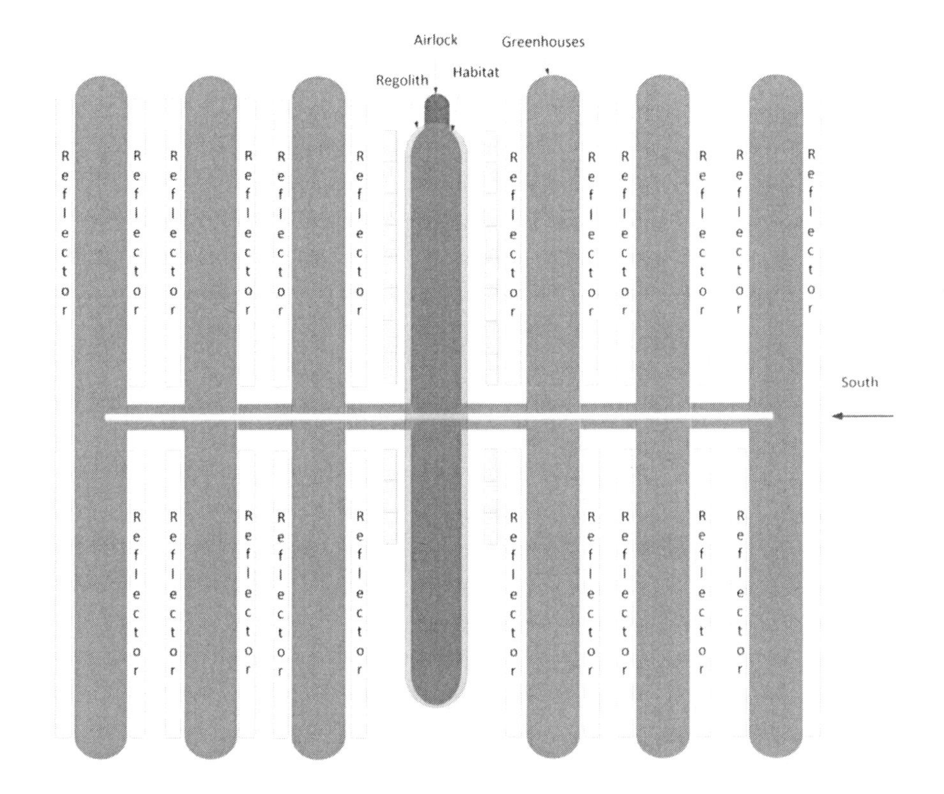

Figure 43: Option 2- Concept for Community. It is easier to assemble and makes reflectors effective.

This is the option we are choosing.

This concept as shown actually has more floor space in the habitat, but is not suitable to build multiple stories with a center height of 4 meters (13 ft). Therefore the internal structure for a second floor is not included. The habitat as shown is 8 meters diameter by 90 meters long and half buried as all previously described habitat structures are.

This option also has a few challenges in assembly but getting regolith on the roof of the habitat is only 4 meters (13 ft) up instead of 12.5 meters (41 ft) up. Even 4 meters is probably above the small backhoe's reach but the backhoe could move most of the regolith with the final cap either added manually with a shovel or possibly a small conveyor belt. (You can walk on the roof when inflated in a space suit but may not want to on top of regolith due to the chance of sharp edges). Because of the issue of piling regolith, we are going to select option 2 for mission planning. It also has the benefit of making reflectors to add sunlight and warmth for the plants far simpler.

Something that could be done also is to connect communities by putting one more tunnel between an outer greenhouse and the outer greenhouse in the next community, possibly with a pressure door. It would allow visiting your neighbors or sharing without having to put on a spacesuit and go through two airlocks. It also provides a secondary exit. The white strip down the middle of the tunnels is intended to be an open path between greenhouses and the habitat to make travel and harvest easy. It could easily be extended to the end if the outer greenhouse is connected to a neighbor.

Balancing the Environments

All enclosed areas will be a sealed, closed ecosystem. We will have to add atmosphere and quite a bit of moisture when we build it, but after that it is retained and recaptured after it is consumed. We will have dehumidifiers that recapture moisture and recycle it for drinking water. We will have some kind of system that recycles waste to water and fertilize the crops. The existence of the plants in the greenhouses will offset the human use of 02 through photosynthesis turning CO2 back into oxygen. In other words, the greenhouses are our oxygen regeneration system.

The formula for photosynthesis is:

$$\mathbf{6CO_2 + 12H_2O + light \rightarrow C_6H_{12}O_6 + 6O_2 + 6H_2O}$$

The net effect on the atmosphere is the plant turns some CO2 into an equal amount by volume of O2. It also produces glucose, which is a sugar both the plant and us can derive energy from and it absorbs some of the water. Of course, when we eat them or the plant dies, the water and nutrients are recycled back into the system.

This is our oxygen recycling system. The acreage suggested for our greenhouses is enough to feed 14 people.

Using Earth air, which is not nearly as CO2-rich as our environment, I calculate our acreage of corn (no we aren't just eating corn, but there doesn't seem to be data for a balance of crops) I get 18.3 kg/day of oxygen production. Each human needs 0.82 kg/day according to NASA so 18.3 kg is all we need plus about 6.8 kg per day excess. This is based on a 2014 study that found that corn consumes 5g/m2 per day of CO2[31].

Whether our greenhouses will produce at exactly this rate is not known. We are going to have less sunlight but our air will be much more CO2 rich, which encourages photosynthesis. We might produce even more or we might produce less and won't know for sure until we test it. If CO2 is consumed at this rate, the plants will be consuming more CO2 than the humans are generating.

Either way, we can monitor CO2 and O2 levels and tweak it if we need to. To raise CO2, we would just pump in some outside air (a small percentage) to our greenhouses. There is nothing wrong with the outside air as long as it is added to our oxygen rich air so we can breathe. It is CO2 with a small amount of argon, nitrogen, and oxygen.

If our over-production of O2 is just like the earth-based measurements at 6.8 kg / day of O2 excess and 9.2 kg /day net of CO2 removed, we then pump in this amount of CO2 periodically to replace what was lost. The net of 6.8 kg O2 would raise our overall pressure 8.2 mbar (2.4%) per month. That is not a short-term problem but would require us to bottle some of our air periodically and provide it for use in new habitats (or sell it depending on how the economy functions on Mars). However, we would be slowly depleting our inert gases doing this so we might have to add some back in eventually. Moving very far toward pure O2 is not safe as far as fire risk.

If, on the other hand, our plants produce a lot less oxygen and we are unbalanced in the other direction, that is OK. The fuel production for a single ship is creating an excess of 63.5 kg/day of oxygen. That is more than 5 times our entire habitat's consumption, and what we might need if there is a slight imbalance in the direction of needing more oxygen would probably only be 0-2 kg per day. However, we would need to remove CO2 eventually which might require a Sabatier / Electrolysis system like we have on the ship. For this reason we might even choose to start at 800 PPM not 1000 PPM CO2 or even much lower in the very beginning. That would give us some margin, especially until the plants have a chance to grow.

Corn is a very large plant. Many of the other plants are very small and should consume less CO2. Also fertilizer such as our waste can be a net CO2 emitter. My guess is we will be much closer to balance than these calculations with just corn would lead us to believe. Either way, we have what we need to adjust as described above.

One other point we haven't touched on is that O2 is being produced in some areas and CO2 in another area. It will not require a lot of flow, but we will want to do some form of air exchange between the greenhouses and the habitat so the plants and people are not choking on their own emissions. This could be a very small blower system piping air from the greenhouses to the habitat for constant fresh, oxygen-rich air, as in a trickle of a flow. The reverse flow would occur naturally to replace the air removed. The imbalance would be slow to develop but can be avoided with a tiny pump and a hose.

This closed system should be very sustainable. The only losses will be trace gases when we go out through an airlock, but that will be designed to pull a pretty hard vacuum, capturing the air and moisture in the airlock before ever opening the door to the outside, so the losses are in the noise especially if we are in the mode of needing to remove air.

Mass

Option 1: Spherical

Table 20 (on next page) shows the total mass to house and feed 14 people using option 1. It assumes the spherical configuration as described and depicted with 4 greenhouses and 1 habitat that is two floors.

Masses	**Quan**	**Each(kg)**	**Total(ton)**	**Description**
Greenhouse				
Shell	4.0	3020.0	12.1	40M Diameter
Relectors	160.0	10.0	1.6	2.5M x 5M Mylar + Frame
Tunnels	4.0	281.0	1.1	4M Dia x 7M w/ support
Equip	12.0	100.0	1.2	Consenser, farm equip
Total			**16.0**	
Habitat				
Shell	1.0	460.0	0.5	25M Diameter
Flooring	491.0	1.0	0.5	
Plumbing & Electrical	8.0	30.0	0.2	
Modular Interior	1.0	4000.0	4.0	
Furniture	8.0	50.0	0.4	
Equip	6.0	100.0	0.6	Consenser, water purify,etc
Food prep and canning	1.0	500.0	0.5	
Airlocks	1.0	1100.0	1.1	Exit
Pressure Doors	4.0	400.0	1.6	On Tunnels
Total			9.4	
Solar				
Solar Panels	216.0	3.0	0.6	1.5 kW/ person
Battery Backups	4.0	480.0	1.9	300 kWHr
Total			**2.6**	
		Total	**28.6**	

Table20: Option 1 – Total mass for spherical configuration

Option 2: Cylindrical

Table 21 shows the mass of option 2 configuration. It is only slightly heavier and still well within the capability of the Starship to deliver to Mars.

Masses	**Quan**	**Each(kg)**	**Total(ton)**	**Description**
Greenhouse				
Shell	6	2709	16.3	8.5M x 107M
Relectors	240	10	2.4	2.5M x 5M Mylar + Frame
Tunnel	6	113	0.7	4M Dia x 9M long
Equip	12	100	1.2	Condensors, farm equip
Total			**20.5**	
Habitat				
Shell	1	823	0.8	8M x 90M
Flooring	706	1	0.7	
Plimbing & Electrical	8	30	0.2	
Furniture	8	50	0.4	
Equip	6	100	0.6	Condensor, water purify, etc
Food prep and caning	1	500	0.5	
Airlocks	1	1100	1.1	Exit
Pressure Doors	2	400	0.8	On tunnels
Total			**5.2**	
Solar				
Solar Panels	216	3	0.6	105 kW/person
Battery Backup	4	480	1.9	300kWHr
Total			**2.6**	
		Total	**28.3**	

Table 21: Option 2 – Mass for cylindrical configuration

The total is 28 tons for option 1 or 28.3 tons for option 2 for 14 people, or 2 tons per person. This does not include consumables we might bring for them or personal effects, just the structure and support equipment, including power.

We did not discuss solar power or battery backup yet, but it will be needed. We have to power all of this. According to references we probably don't have to heat the greenhouses. We might have to heat the habitat, but if so it will be very limited with it buried under over a meter of regolith. There will also be some equipment to run and lights. The power listed is 21 kW continuous including nighttime as required or 1.5 kW per person. (The peak output is actually about 66 kW at solar noon but we are also recharging batteries during the day and the power output varies with Sun angle).

As discussed in the habitat section, it is believed the cylindrical version (option 2) has far fewer assembly issues and the benefit of making reflectors work effectively. Therefore, we will assume option 2 for our mission planning.

Vehicles

In the scenario that follows, we are going to want to use vehicles. One is a covered rover. Another is a trailer for moving equipment around. We will also need heavy equipment for moving regolith and for building our habitats and greenhouses.

The rover would probably be similar to the vehicle in the movie *The Martian*, but much smaller. That vehicle was massive. We wouldn't ever be able to get it into or out of a Starship, as massive as the Starship is.

Starship will have a hatch on the lowest level of the cargo area of the vehicle, which is about 30 meters (100 feet) up. I am also assuming the hatch is an airlock. It is not practical to have an airlock from one deck to another. The floor of the pressurized area would have to be curved (wasted volume) or massively strong, as in massively heavy. It would make more sense to have an airlock at the hatch that is about 2–2.5 meters tall and 2–2.5 meters wide in order to get vehicles, equipment, and personnel out. This also drives vehicle size.

The rover and other vehicles need to be on the bottom deck and a wench would be required to get them and all of the other equipment down to the surface, including a platform for personnel and smaller items that we don't want to have to wench down one at a time. A wench will also be required inside the Starship to move cargo from higher decks to the lower deck for removal from the Starship and a third one to get things off the trailer at their final location. Because of the huge volume of cargo (about 4 decks worth), it is not practical to move wenches back and forth every time we need to unload a trailer.

We want the rover to be pressurized. This means it has a strong outer shell and the capability to pump the air out so we can enter and leave on Mars' surface. We don't have to pressurize it for short trips. If we are going 500 meters we probably do not pressurize it. However, if we are going 148 kilometers as described in 'The Trail' previously, we pressurize it and take our helmets off at a minimum. The rover also has to have some level of CO2 capture for long trips. Perhaps not a whole recycling system, but you can't have the humans choking on CO2 after a few hours.

The length is under 3 meters. We have an airlock on that lower deck on the Starship. It has to drive into the airlock and the door has to be closed behind it.

All of this has to happen within 8 meters diameter and the airlock has door(s) that open, hopefully double doors to limit door length.

The dimensional limits become somewhat obvious in Figure 44, which depicts what the loading on that deck might look like. It is not completely volumetrically (3 dimensionally) optimized but you probably get the idea. More vehicles or bigger vehicles become a challenge.

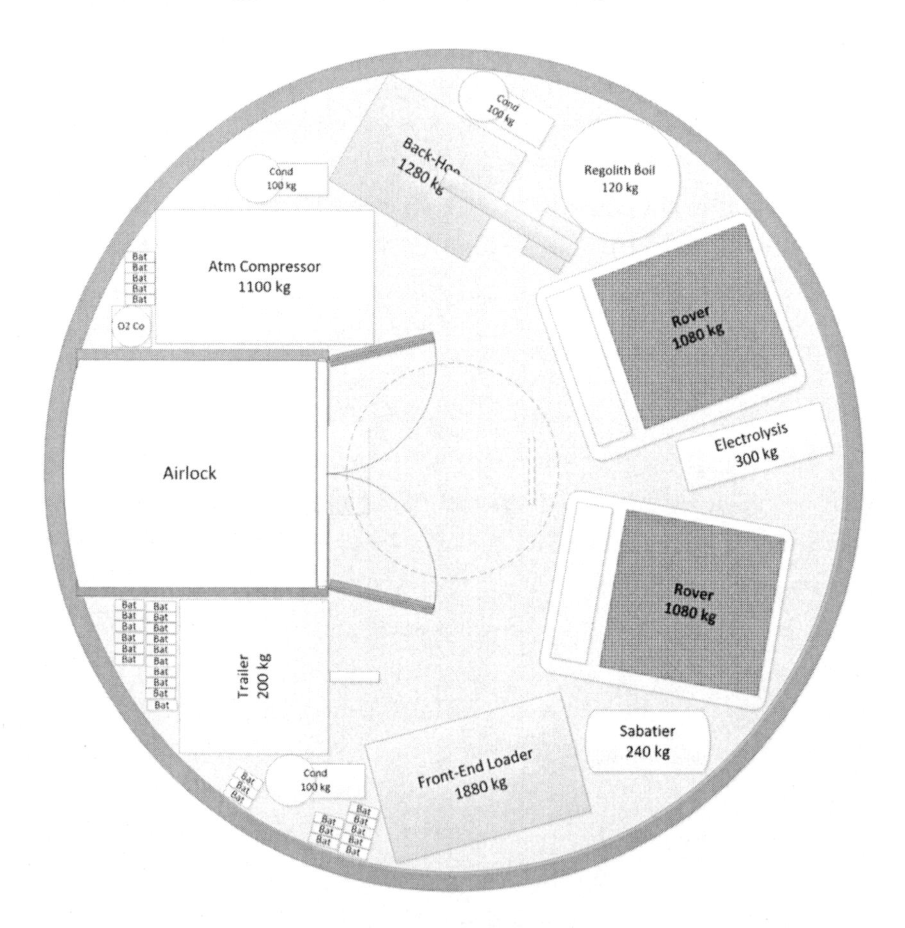

Figure 44: An example of space available on the first deck and why a vehicle must be small

The Rover is depicted as 2.4 M (7.9 ft) long and 1.9 M (6.2 ft) wide, with solar panels on the roof. As you can see, this may be about the limit on size. The center circle is under the transfer tube, where the crew comes down. It is sized at about 2.1 meters internally so other cargo can be removed from higher decks. The white thing within this area is intended to be a retractable ladder to come down on Mars. (It is not required in zero-g.) The rover would have to drive into the airlock and the doors be closed behind it in order to get out.

The rover would have an electric motor and be powered by a 75 kWH battery. This would give it a significant range.

Having the rover gives the crew the mobility to explore on early missions. It would also be used to move equipment around using the trailer that will be discussed next.

The trailer in the figure is in order to move equipment. First of all, we have over 11,000 m2 of solar panels and their frames that have to be moved and arranged over the area of a few football fields. We also have a lot of equipment that needs to be set up on the Martian surface for fuel production. This is probably a simple flatbed trailer, perhaps with some low rails just to prevent equipment from sliding off. We would also need some kind of wench apparatus to get things on and off the trailer. A lot of the equipment is too heavy for the crew to move unaided.

If we found an even better water site several kilometers away, we would want to move the equipment after we fuel the ship but before we leave. That would also require the rover and trailer.

We need heavy equipment. A small backhoe and a front-end loader might be the best options. We need it to collect regolith for water extraction, lots of regolith, like a ton and a half per day assuming 50% water ice. We may also need to move regolith for other reasons, like putting our habitats and greenhouses in place. The vehicles would be electric and powered by perhaps one 75 kW battery each. They would use power fairly quickly as in use a battery charge in 3 to 4 hours but the regolith movement is probably only 1 or 2 times a day, so they should only need charging every day or two for this operation, but perhaps more while we are digging out for our greenhouses and habitat. The reason for suggesting both is that the backhoe is better for digging and the front-end loader is better for scooping and moving large quantities.

The size depicted in the previous graphic was 1.5M (5 ft) wide by 2 M (6.5 ft) long. That is a very small vehicle. That makes the design challenging to actually have a scoop on the front end with that wheel base, and still be as light as possible. It might mean a very small scoop in order to not get off balance, in addition to putting most of the weight in the very back and perhaps shipping it with the scoop up as long as that isn't above about 2 meters.

The estimated mass of these vehicles is shown in Table 22 below. This is based on one 75 kWH battery each for the powered vehicles.

Vehicles				
Item	**Size**	**Veh Mass(kg)**	**Power Mass (kg)**	**Total (kg)**
Closed Rover	2.4M x 1.9M	600	480	1,080
Trailer	1.5M x 1.5M	200	0	200
Front-End Loader	2M x 1.2M	1400	480	1,880
Back-Hoe	2.5M x 1.2M	800	480	1,280

Table 22: Vehicle sizes and masses

Recycling Systems

There will be recycling systems on Mars. The oxygen is probably taken care of by the plants. However, water and waste must be dealt with. A significant part of water recycling is just having dehumidifiers that collect excess atmospheric moisture and provide it for drinking water. That takes care of evaporation from watering plants and human perspiration and respiration losses.

We will also have to deal with human waste. Solids could probably be dehydrated, returning water into the system. The resulting waste would probably be used to provide additional fertilizer for the plants. The soil can always use biological material and nitrogen for fertilizer. Liquid waste might be used directly to water the plants. Studies show urine to be an excellent fertilizer[32].

Over time, there will also be biological waste from the greenhouses, as in parts of plants trimmed off or dying plants. This would all also be used as fertilizer. Feeding the plants is probably the best use of rotting biological materials. It would be things we don't eat but are left over. Everything needs to be recycled the best we can. The soil needs it. Forests have recycled waste that way for millions of years. The Martian soil is not rich in nutrients at all. We will kick-start it with fertilizer but have to keep feeding it with resources after that. With care, over time it will become rich black soil every farmer dreams of, and we will have PhDs in agriculture advising us. Hopefully one or two of them will move to Mars.

The recycling applies to everything. We won't use disposable bags. We won't use disposable wipes. We will strive for materials that can be reused. We will wash out containers and reuse them. We will use reusable wipes like wash clothes and clean them. We will revert to cloth diapers. There won't be any landfills or trash. If we use paper and throw it away, but we probably won't, at a minimum it will be buried in the green house to feed the plants and we don't have space to grow trees to replace paper.

In time, we will have trees, like perhaps a shopping mall that has sunlight and a park in the middle, but not in the early years. If we do, we won't cut limbs off and haul them away. All organic waste goes back to the soil.

If you go to Mars, you won't just be a vegan, you will also be an expert in recycling. We will be in a closed, limited environment. We won't generate trash and we definitely won't use Styrofoam containers.

Our Path to Colonization

The land belongs to the future, Carl; that's the way it seems to me. How many of the names on the county clerk's plat will be there in 50 years? I might as well try to will the sunset over there to my brother's children. We come and go but the land is always here. And the people who love it and understand it are the people who own it—for a little while.

Willa Cather, *O Pioneers!*

First Human Mission – 45 Degrees

I have no control over what SpaceX will do. They may launch unmanned ships to Mars first as a test flight (or not). If they do for proof of concept that is responsible and is pretty conservative but they would be stranding the unmanned ships until humans arrive to set up a fuel factory to refuel them. I will start with human missions. I do not believe an unmanned rover can do what humans can.

The first human mission is probably composed of 2 ships that will land. It will focus on the task of finding the best local water source. They will need 715 kg of water per day for fuel production per ship. However, they intend to eventually supply the southern site unless they find a good water source of their own. The southern sites will need just as much for fueling but will need more. As they start colonization, they will need many tons for purifying soil and for starting up habitats and greenhouses. The northern site needs a very good water source. They can manage with 30 or 40 percent water ice content in soil, but a better source would significantly reduce power requirements and volume requirements.

Both ships will carry their own fuel factory. Since each ship needs a couple of football field size area for solar cells, the ships will probably land a hundred or two-hundred meters from each other.

Besides providing payload capability, two ships create a level of safety where all of the crew could get back on one ship if they had to. The extra ship is intended to be used to fly back but is also something of a lifeboat.

SpaceX could also choose to send a third ship that is an unmanned orbiter. You don't want to leave humans in orbit for a year and a half with the radiation levels in space or the zero-g health issues. It could provide imaging, mapping, navigation, and other support but also act as a lifeboat. If the ships on the ground had issues with water quality, as in percent content driving huge increases in power and volume of regolith required or if they had issues in production for other reasons, it would leave an option. If they could just get one ship fueled with 198 tons, instead of 732 tons each, they could get that ship to orbit and transfer to the ship already in orbit which left Earth orbit with enough fuel to return. Earth orbit to Mars orbit and return is only about a 71% fuel load. However, Earth orbit to Mars ground and back is out of reach without refueling.

The human ships are taking enough food to last an additional 2 years and 49 days until the next launch window so the orbiter isn't necessary. It is just a choice that could be made by mission planners for the first mission in order to assure getting the crew back on time.

The first human mission obviously has to carry the fuel factories, one per ship, including water production. That mass is shown in Table 23 along with volume, which is important for determining how well we can fit it in. That will be shown only for this mission. It will be similar for future missions but the point is it will usually take most of the ship. Early missions will include their own fuel factory. However, these fuel factories are a valuable asset that can be used over and over.

Fuel Production Mass Requirements:				
Item	**Quantity**	**Mass Per(kg each)**	**Total (tonnes)**	**Vol (m3)**
Solar Panels	11400	1	11.40	57.00
Solar Frames	2850	8	22.80	28.50
CO2 Tank	1	126	0.13	1.52
H2 Tank	4	544	2.18	14.32
02 Tank	8	358	2.86	17.85
Water Tank	1	137	0.14	3.58
Sabatier	1	600	0.60	0.14
Elecrolysis	1	300	0.30	0.14
Atm Compr	1	1100	1.10	5.10
H2 Compr	1	300	0.30	1.50
02 Compr	1	40	0.04	0.04
Liquid Pumps	3	30	0.09	0.12
Cryo Pumps	2	100	0.20	0.40
H2O Condender (Sab)	1	250	0.25	1.25
CO2 Condenser(Sab)	1	100	0.10	0.50
CH4 Condenser	1	400	0.40	2.00
H2O Condenser(O2)	1	100	0.10	0.50
02 Condensor	1	600	0.60	3.00
Plumbing	1	80	0.08	0.20
Radiarors	65	13.5	0.88	1.65
Batteries	7	480	3.36	0.56
Umbilicals	4	100	0.40	0.20
Regolith Boil	1	2175	2.17	1.50
Structure and Margin	1	1000	100	1.00
		Total	**51.5**	**143**

Table23: Mass required for fuel production at 45 degrees (with water capture)

This is 51.5 tons. One per ship is required but we need a lot of other things too. We have crew and supplies. We might want to take extra supplies in case we have issues and have to stay another 2 year plus cycle. We want heavy equipment for moving dirt around, including collecting regolith for water production.

A key goal of the mission is establishing the refuel capability and a northern base. We will also search for better water sources such as nearly pure ice or even an underground stream. This means scientific equipment and mobility. We definitely want rovers for the crew. We will want the capability to search for water underground, and the capability to drill if we actually find a deep underground water source.

A suggested payload is shown in Table 24.

Payload	Number	Mass(tonnes)	Total (tonnes)	Item	Vol (M3)
Crew and Provisions	8	2.37	19.0		28.4
Extra Provisions	8	1.35	10.8	Food to survive an extra cycle (2 years 49 days) if needed	16.2
EVA Suits	8	0.05	0.4		1.2
Fuel Production	1	51.5	51.5		143
Extra Solar Panels	700	0.003	2.1		7
Closed Rover	2	1.08	2.2	Explore site for best water source, move equipment	12
Front-End Loader	1	1.88	1.9	Be able to dig and move dirt around	4
Back-Hoe	1	1.28	1.3	Include a jack hammer and ice saw	4
Ice Breaking Equip	4	0.1	0.4	To move equipment	0.2
Trailer	2	0.2	0.4	Sample for soil content, search for water, etc	1
Scientific Equipment	20	0.2	4.0	Drill for water	2
Bore	2	2	4.0		0.4
Well Equip	1	2	2.0	Plexiglass geodesic dome unpressurized(20M dia)	0.5
Enclosure	1	0.6	0.6	The payload is 100 feet up in the air and on multiple decks	0.6
Wenchs & Platform	3	0.5	1.5		0.375
Spares/ Tools	1	8	8.0		8.0
Margin/ Other Equipment	1	10	10.0		10
Total			**120.0**		**239**

Table 24: Possible payload on first human mission

The extra provisions are a couple of years of food, some water, and necessities in case we don't get refueled in time and have to wait for the next window in 2 years and 49 days. This is a basic requirement for survival if we have any issues with fuel production or water. If we don't use it we will leave it for the next crew. The extra supplies on each mission could also include meat, perhaps lots of it as additional stockpiles are added and it could be consumed. So perhaps, Martians won't be completely vegans, at least in the early days if we stockpile meat. Eventually, they may have their own sources.

We need heavy equipment to collect regolith. This payload and the prior discussion suggest a front-end loader and a backhoe to uncover the moist regolith and then dig it up and carry it to the regolith boil equipment. The payload also includes equipment to break up ice, like an ice saw and perhaps a jack-hammer. We may also want to scoop up in other areas to search for high ice content.

The enclosure is an unpressurized Plexiglas geodesic dome 20 M in diameter. It is intended to provide UV and dust protection to equipment and supplies left behind so they are not ruined or inside a sand dune when the next ships arrive. All of these items will be left under the dome for the next mission except for crew and supplies for the return trip. The next mission arrives about 229 Earth days after this crew leaves.

The purpose of the first mission is proof of concept, locating the best local source of water, and to leave the fuel production capability on Mars for future missions. It will be a lot of mass future missions don't need to carry. The crew will collect moist regolith and boil water out of it. If they are really lucky they will boil huge chunks of dirty ice. They will use this and Martian air to produce the fuel for their return.

The margin might be used for a lot of things. One option is more spares or solar panels. However, there is certain to be equipment SpaceX wants to take that is missing from this list.

The fuel production includes preloading all of the tanks a bit. Each will be at least at atmospheric pressure. Why not with the commodity it is intended to hold, except that water might be water vapor and oxygen. Liquid water is too heavy. Maybe we pressurize some to more than one atmosphere. For example, the hydrogen tanks only hold 36 kilograms of hydrogen each at 120 bars (1764 psia), due to its extremely low density. Why not load each one to its max capacity and save just a little on electrolysis and compressor power.

Figure 45 shows an approximate internal use of Starship, the cargo is about 239 m3, which is over 4 decks based on 60% efficiency in storage above the first deck. This includes an allowance for an inefficient first deck loading with drive-able vehicles and also an efficiency allowance for things like spherical tanks and a lot of square solar panels in a round vehicle, so it is probably floor to ceiling in decks 2 through 4. Note that the external diameter of Starship is 9 meters. However, the internal diameter for payload is only 8 meters according to SpaceX. It is believed this is because there is plumbing, wiring, etc. between the external and internal walls, but might include the long payload hatch shown on their website, which would not be part of a crewed ship, so we might be a little better off than the numbers used here.

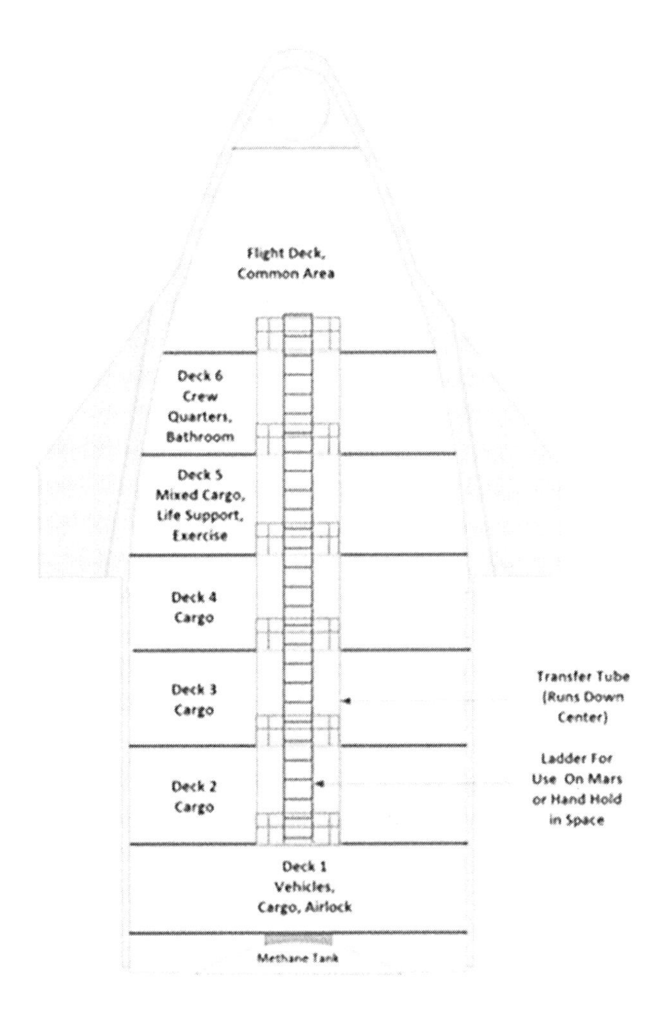

Figure 45: Vehicle Internal Usage

The assumed deck height is about 2.1 meters (6 ft 11 in), making the volume per deck 97 m3 after allowing for the transfer tube. Any change in the height would simply change how high things are stacked and how many decks there are. This drawing was laid over the actual cargo bay cross-section to scale, but the cargo bay cross-section was not included due to copyright issues. Those dimensions are available on the SpaceX website as the user's guide for Starship.

This usage is very different from most artist concepts. Starship is a large volume but the fuel production is very large too. The majority of the fuel production is just solar panels assuming a thickness of only 0.5 cm. Also, things like rovers and heavy equipment use a lot of space and are inefficient because they turn the first deck into a parking garage, though they may stack tanks and other items like that on top of them and unload those first.

The artist renderings, like Figure 5 back in chapter 3, might be closer for missions like a tourist ride to low Earth orbit rather than a mission to Mars. A mission to Mars will have massive cargo.

There are widely published external views for a crewed ship, with windows presumably in crew quarters on one side and continuing up through the observation deck that start somewhere near the top of deck 4. Those may be roughly the actual external view of a Mars bound ship or the window area might be reduced. These ships will be very heavy in cargo, limiting crew space. Also, there won't be a whole deck dedicated to exercise equipment like many renderings. That would be very under-utilized with a crew of 8.

When the crew comes back, the return ship will not take any of the cargo back to Earth. It will just be the crew, supplies for the return trip, and perhaps some Mars samples for analysis on Earth. The return payload will only total about 4 tons.

Second Human Mission – 45 Degrees

The second mission will also probably be two ships. The crew per ship will again be about 6, perhaps 8. There is no reason for huge crews, but a lot of work needs to be accomplished so two or three crew members probably doesn't work.

This mission will bring with it the charging stations for the trail to the south. It will also bring the vehicles for water transport, at least one per ship so they can run like a shuttle service. It will bring the regolith boil-off, condenser, and water tanks for additional water production to support the southern site that will land next time.

This mission may be at a new site. Based on that fuel production was included but they may be able to grab the fuel production from the previous mission and use that or at least add it to their resources and spares. The goal of the first mission was in part to find the best water source in the area, so presumably this mission has an excellent water source. That might be 60% water ice content or even better, buried sheet ice or the mother lode, a geothermal well. They will have significant and growing water requirements, not only their own but supplying the southern site. The plan includes providing the southern site water before they even arrive.

The probable payload for the second mission is listed in Table 25.

Payload	**Number**	**Mass(tonnes)**	**Total (tonnes)**
Crew and Provisions	6	2.37	14.2
Extra Provisions	6	1.35	8.1
EVA Suits	6	0.05	0.3
Fuel Production	1	51.5	51.5
Front-End Loader	1	1.88	1.9
Back-Hoe	1	1.28	1.3
Solar Panels/ Depot	5.5	2.144	11.8
Rig for H2O Transport	1.5	2.5	3.8
Extra H2O Trailers & Tanks	6	0.355	2.1
Regolith Boils	1	2.2	2.2
Extra Solar Panels	1600	0..003	4.8
Extra H2O Tanks	2	0.14	0.3
Extra O2 Tanks	8	0.4	2.9
Habitat (8m diameter x 48m)	0.5	3.0	1.5
Wench	4	0.5	2.0
Margin/ Other Equipment	1	10	10.0
Total			**118.6**
Per Ship			
Focus of mission is building infrastructure for water transport			

Table 25: Payload for the second human mission

In the payload, a habitat was included. That is probably a smaller 48 meter by 8 meter diameter cylindrical version of the habitats described in the previous section. It would include some basic life support like the systems described back in chapter 4, including a Sabatier reactor for CO_2 recapture, electrolysis, and condensers to recycle water. It could be used by the crew and could be a refuge for future missions or for travelers years down the road. The mass includes some interior structure and furniture, as well as these systems.

Constructing a habitat is also a proof of concept. It demonstrates the capability to construct this on Mars with limited assets in space suits. There will be an assembly kit included. One obvious issue is supporting the ceiling while filling it half way with regolith. That might involve a pole structure, like for a tent. It is also unlikely they would drive the heavy equipment into it, at least not until there is a significant layer of regolith to protect the shell. The importance of getting it right grows with each mission. This mission could live without it, but for future missions it is more critical.

The primary focus is putting the water transport in place so that the northern base can provide water to the southern base. Hopefully, delivering water will not be necessary forever, but it is until the southern location finds its own water source, which is probably not 50% ice regolith like is available at the 45-degree location.

The mission also includes extra water transport capability. This is so that once they have the charging stations in place they can pre-deliver up to 45 days of water for the first southern mission to avoid delays in fuel production at the southern site. Since the water tanks will be insulated and have solar panels on their roof for heat as required the water can be kept liquid until the next human mission arrives at fifteen degrees latitude.

Installing the charging stations will be quite a task. It will involve two crew members in a rover pulling a trailer with all of the equipment. The first trip might travel about 6–8 hours to the first location and set it up. This might take one day or more, depending on how simple it is to assemble. The crew would probably sleep in their rover and recharge their battery before returning.

For the second station, they would drive to the first station and recharge there. They might drive on a couple of hours but it will be getting dark soon. They sleep in their rover and then arrive at the second location the next day and assemble it. Before they leave the third day they must recharge. They drive 6–8 hours to the first station and recharge again. Then on the fourth day they drive back to the ship and recharge their vehicle again. It would build in a leap-frog way like this.

Fairly early in the process, a second crew would join the activity so that one team puts in station 3 and the other puts in station 4 and the trips will get longer and longer. If a 24-hour a day automated vehicle takes 6 days one way, obviously the round trip could be 3 weeks or more by the end.

One important issue would be to charge the batteries they will transport each time before they leave, or before they leave Earth, so they can recharge immediately. Otherwise it will take them most of a day for the solar cells to recharge their rover. Each depot is sized for a little over one charge per day on average at about 80–95 kWH per day, depending on sun angle. The crew will set the solar panels to an angle that minimizes this variation and then leave them that way forever, with no additional adjustments since the depot is unmanned.

The depots can always be sized up if needed. Adding more solar panels increases capacity. Having two batteries provides it some buffer to tolerate two vehicles arriving the same day due to scheduling. However, more batteries could be added if the northern location were providing water for a large numbers of ships in the southern location.

The southern location will hope to locate and drill for water. When they do, the northern location may go away, but that could be several years. Even if the northern location is abandoned, the trail might be left in place allowing those at the southern location to travel north if needed for additional water or other resources, even if it is just to plunder the fuel production and solar cells at the northern base. Anyway, who would want to destroy a trail/highway that can be easily traveled by new citizens of the planet Mars.

Figure 46 shows a possible crew deck layout. A walkway is provided so that the crew can get around on Mars. In zero-g it would not matter. The crew quarters would be big enough to allow a bit of privacy and comfort. Each cabin would have a door from the transfer tube. Minor details are shown in order to communicate the concept.

Obviously, I assumed couples based on the figure. I would think on a two-and-a-half-year mission, it would be better to travel with your significant other rather than leave them behind. If singles choose to go, the quarters could be subdivided or they could just share it.

My intent is not exclusion at all. I said couples. I did not say heterosexual couples. I could not care less what gender or preference the couples are. I just think two and a half years is a very long time to be alone. I wouldn't want to leave my wife behind. The diagram was sized for 5 couples. This could be adjusted or rearranged as needed. The point is the concept of private quarters. Future missions may have ten people or more. The crew quarters could also expand into the deck below if space allows.

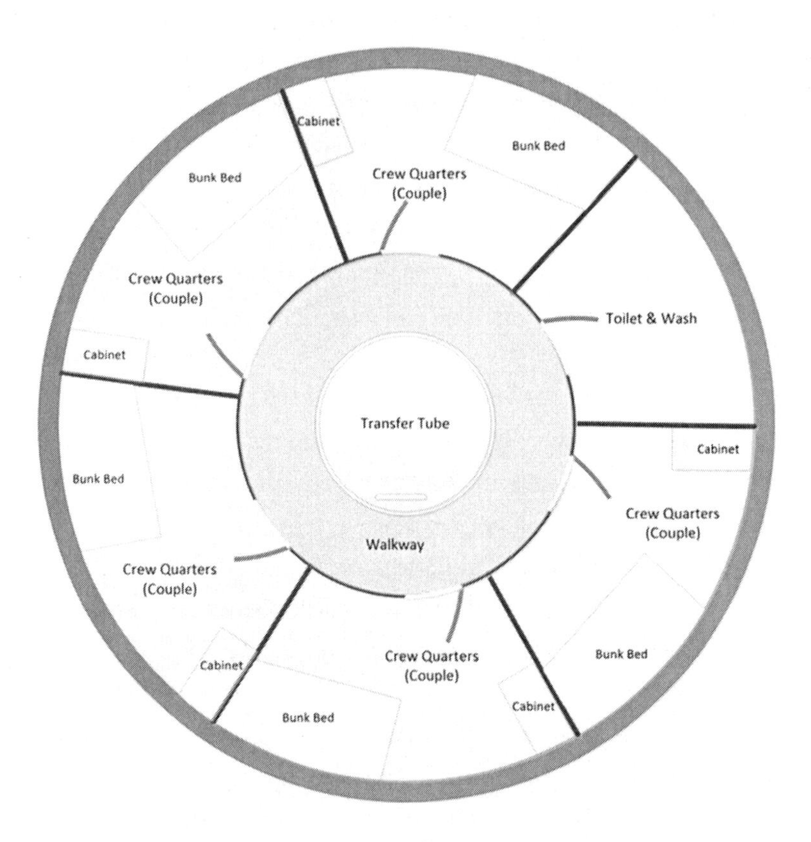

Figure 46: Possible crew deck layout for a crew of ten

When the crew comes back, the return ship will not take any of the cargo back to Earth. It will just be the crew, supplies for the return trip, and perhaps some Mars samples for analysis on Earth. The return payload will only total about 4 tons.

Third Human Mission – Going South

The first mission to the southern, tropical location will probably have a crew of about 8 per ship and include 2 ships. The goal is to locate the best possible location, not to start colonization.

This mission will start the southern presence at 15 degrees. Ships will land at both sites, but the northern missions are more crew change out than doing anything new, other than possibly providing additional supplies or more water generation and transport equipment. From here on, our focus will be the southern missions. These will lead to colonization.

This mission will bring fuel production to the southern location. It will also bring in its own heavy equipment to start construction of the base that will be there, scientific equipment, and enclosed rovers for exploration. The suggested payload is shown in Table 26.

The southern base will also bring the capability for water production. Whether this is producing small quantities from the 3 or 4 percent in the top layers of soil or finding a much better source is unknown but the hope is definitely there to produce their own water.

A significant part of the mission goal is to find an optimal site for the permanent southern base. There could be an underground geothermal site. That would be an enormous asset. It would mean not only water, but also power. This mission is both a first at the lower latitude, but also a reconnaissance mission. The crew will have rovers and scientific equipment to find underground water. A good outcome would be water ice a couple of meters down. A great outcome would be a geothermal spring underground. Either way, the mission will search for their future water source.

Payload	**Number**	**Mass(tonnes)**	**Total(tonnes)**	**Remarks**
Crew and Provisions	8	2.37	19.0	
Extra Provisions	8	1.35	10.8	Food to survive an extra cycle(2 years 49 days) if needed
EVA Suits	8	0.05	0.4	
Fuel Production	1	51.5	51.5	Take regolith just in case.Wort case we could produce a little at 3 or 4% ice
Front-End Loader	1	1.88	1.9	Be able to dig and move dirt around
Back Hoe	1	1.28	1.3	Be able to dig and move dirt around
Enclosed Rover	2	1.08	2.2	Mobility to search
Trailer	4	0.20	0.8	Possible movement of equipment

Additional Solar Panels	800	0.003	2.4	Can take on rovers for longer trips or supplemental power production
Scientific Equipment	2	4.00	8.0	Sample for soil content. Search for water, etc
Bore	2	1.00	2.0	
Well Equipment	1	2.00	2.0	
Small Greenhouse, Seeds, Fertilizer	1	3.00	3.0	Experiment with plant growth
Supplies for greenhouse	1	1.50	1.5	Could include atmosphere, tools,etc
Extra H2O Tanks	4	0.14	0.5	
Extra O2 Tanks	8	0.36	2.9	
Enclosure	1	0.60	0.6	Plexiglass geodesic dome unpressurized (20M dia)
Misc/ Spares/Tools	1	8.00	8.0	
Total			**119**	
Per ship				
Focus of mission is Determining Best Tropical Base Location				

Table 26: Mission 3 – Southern payload

We would like for the location chosen to be the site of a future city, though cities on Mars will include a lot of farmland. Until the population becomes so diverse as to support farmers and city people, everyone, or at least some people in every commune will be a farmer.

A small greenhouse is included in the payload, perhaps 8x24 meters with an airlock. The idea is this crew will bury and inflate the greenhouse as described in chapter 8 and plant crops. The crops will not be sufficient to feed everyone but the point is we have never grown plants on Mars. This would be a demonstration before leaving people there on the next mission and also a learning experience as to what grows well and doesn't in the Martian soil enhanced by fertilizer and the crew's waste.

The southern site will be the first of many that require more water than just for fuel production. Water is required to purify the soil in the greenhouse and also to start it. If the crew use their waste to fertilize it that is also water that is being lost from their ship environment. While they can produce at least some on their own, this will require the northern site to provide a steady supply. The trail will see continual use bringing water south.

Fourth Human Mission – The Real Start of Colonization

The fourth human mission will expand to 4 ships at the southern location. Each ship will take a crew of twelve. Ten of those will stay behind and two are corporate employees that will help and guide the colonists for the first year and a half but will go back with the ship. Or perhaps the corporate employees wish to stay also. Either way, one ship will be left behind as a refuge in case it is needed. While no return crew is there it may be fueled with the equipment left behind just in case or for a head start on the next mission's refueling.

It is believed that at least by the fourth mission, with one every 2 years and 49 days, colonization will really begin with the introduction of greenhouses and habitats. These greenhouses will provide the food for inhabitants, so they no longer have to rely on provisions brought from Earth. It is probable that on the previous mission there will be at least experimentation with growing plants on Mars but this brings it to a whole different scale. The colonists will establish habitats and greenhouses to house and feed themselves. This will be the early beginnings of a city.

These early inhabitants will be learning how to live on Mars but they will also be paving the way. They will grow their own food. This will not be an easy life but perhaps a rewarding one. My grandparents were farmers. One grandfather was asked what he would want more than anything. His response was to live a thousand years where he was with my grandmother. How many people do you know that would respond that way? Hard work, but self-reliance, freedom, and satisfaction is not that bad. He also lived to 98, so there is something to be said for hard work and a happy life.

As described in chapter 8, these people will grow their own food and then can it or freeze it to make it last until the next crop. Farmers have done this for hundreds, if not thousands of years on Earth. Well, maybe not freezing it for that long but canning has been around for a while and potatoes can last months in a cool dark place. That is why people had root cellars.

The suggested payload is listed in Table 27.

Payload	Number	Mass(tonnes)	Total(tonnes)
Crew and Provisions	12	2.4	28.44
Extra Provisions	12	1.4	16.2
EVA Suits	12	0.1	0.6
Fuel Production	0.5	51.5	26
Greenhouse & Habitats	1	27.6	27.6
Fertilizer	1	4.0	4
Front-End Loader	1.5	1.88	2.8
Back Hoe	1.5	1.28	1.9
Enclosed Rover	1	1.08	1.1
Infrastructure to assemble Greenhouse and Habitat	1	2.00	2.0
Extra Solar Panels	700	0.003	2.1
Extra O2 Tanks	8	0.36	2.9
Extra H2O Tanks	2	0.14	0.3
Scientific Equipment	1	1.0	1.0
Misc/ Spares/Tools	1	2.0	2.0
Total			**119**
Per Ship			
Goal is Starting the colonization of Mars			

Table 27: Southern payload for human mission 4

This payload assumes 4 ships this time. The 0.5 fuel production is because we have 2 and need 2 more for the four ships. The table is just balancing payload. It is probable that two ships will bring their own fuel production and the two that already have it will bring the greenhouses. The table is intended to simply capture the total payload and show the equivalent per ship.

We are now in this to stay. We are bringing habitats and greenhouses. We are going to leave people on Mars. We are starting a base / colony. We may be still searching for water but believe we are in the best location possible in the area. We can build a small town here and continue to look for the geothermal source if we have to. Perhaps we have found it already. We brought a bore to drill a well on the previous mission. When we eventually do we may shut down the northern site, but until we do it is providing us with a critical resource, water. Our growth however is limited by their growth. Until we find water they have to produce it for each ship and also for use in the greenhouses. As we discussed, purification of the soil, plus just watering and hydrating a large environment will require additional water.

As other missions follow on, they will bring more habitats and greenhouses and more demand for resources. Hopefully, we find water after a mission or two. If not the northern location has to grow with us to provide water. We will also require more power than just local communes generating their own and no community resources. When we get to the size of cities we need a lot more than just independent survivable enclosures. We eventually need utilities and shared resources like trade centers or markets. We also need a power grid where communes can take from it or feed it.

Fifth Human Mission – Growth

By the fifth human mission, we have established feasibility and have 43 permanent residents at the southern location. The three is because there were three births on Mars. It is time to continue forward with colonization. This mission will include 6 ships, with an average crew of 15 each. Again, 2 crew members will go home with the ship but 13 per ship average are going to stay. They brought their own habitats and greenhouses. They will be additional colonists. They have extra provisions in case they need them while getting their greenhouses up and running, or perhaps they will have children while on Mars and need a little bit extra.

The suggested cargo is listed in Table 28 (on next page).

Payload	Number	Mass(tonnes)	Total(tonnes)
Crew and Provisions	15	2.4	36.0
Extra Provisions	8	1.4	10.8
EVA Suits	15	0.05	0.8
Fuel Production	0.333	51.5	17.2
Greenhouse & Habitats	1.33	27.6	36.8
Fertilizer	1.33	4.0	5.3
Front-End Loader	1.33	1.88	2.5
Back Hoe	1.33	1.28	1.7
Enclosed Rover	1	1.0	1.0
Trailer	1	0.2	0.2
Infrastructure to assemble Greenhouse and Habitat	0.67	2.00	1.3
Extra O2 Tanks	8	0.36	2.9
Scientific Equipment	1	1.00	1.0
Misc/ Spares/Tools	1	2.00	2.0
Total			**119**
Per Ship			
Goal is Continuing the Colonization of Mars			

Table 28: Payload per ship on the fifth human mission, which is the third southern mission

This mission will take the population of Mars from 43 to 121. This kind of growth can be sustained. It is mostly limited by the number of people and resources we can take with us. If we grow the number of ships by about this proportion (4 to 6) it pretty much limits us only by payload. Crew plus the greenhouses and habitat is about 4.9 tons per person or 68.4 tons per colony of 14. We might add some to that in the form of extra provisions, but probably can slow down a little with the heavy equipment since it can be reused (these are not included in the 68.4 tons). We also have to bring enough fuel factories for the extra ships.

With exponential growth, we have to provide atmosphere and water to each new habitat. Each fuel production supplies an excess of about 30 tons of oxygen per ship. A complete habitat with 6 greenhouses requires about 5.7 tons so we are at least supplying this at a steady rate, more than able to keep up with the growth of about 14 people per ship and one habitat and six greenhouses associated with the 14 people.

The next mission, keeping this ratio, might be 9 ships and 12 habitats for an addition of 126 more people. However, we will have to back off on that ratio just a little at some point. People will be born on Mars and we will need additional habitats to accommodate them. This means perhaps only bringing 10 or 12 people per habitat. The idea of course is that everyone is included, but there will be people procreating. This is an ongoing colony that will be here for perhaps millions of years. How that will turn out is hard to say but procreation is a good thing. It is part of how we build a society and keep the species going.

Terraforming

You know what I love most about Mars? They still dream. We gave up. They're an entire culture dedicated to a common goal, working together as one to turn a lifeless rock into a garden. We had a garden...and we paved it.

Quote from *The Expanse* by James Corey

Mars can be colonized as it is, but we may eventually want to try to transform it. This may be many years from now and may take centuries to accomplish. It is possible but it isn't easy. There are a few things required. First the magnetosphere is too weak to hold the atmosphere in effectively. That means air particles are very slowly escaping to space. It was reported that two of the Mars probes detected gases trailing off Mars away from the Sun. That is an issue. If we warm it and create more of an atmosphere in the process, will it be lost to space?

At least one scientist has speculated that Mars lost a lot of its water when it was warmer because the water vapor broke down in the atmosphere and the hydrogen, being the lightest gas, rose and was lost to space. Mars has a lot of frozen CO2 and water. To lose them would be tragic, though it would be a very slow process, taking thousands or even millions of years.

Let's look at the options.

The Magnetosphere

I have read numerous articles on how superconductors could be used to create a current around Mars and therefore a magnetic field. That may be true but I think putting superconductors around the planet (over 21,000 km) is beyond anything we are prepared to do currently. When that day comes, we can protect the atmosphere.

However, if we send thousands of people to Mars, perhaps tens of thousands, they may be the ones with the motivation and desire and capability. My guess would be that they build it in a few generations. They will have access to a lot of resources on Mars and in the asteroids, once they get past their infancy as a colony. They will go through phases of surviving, then thriving, then conquering. Humans are clever and always win out when faced with a challenge if they are educated and motivated.

During conquering, they will mine the asteroids and have more resources than the Earth will. The asteroids contain trillions of dollars' worth of rare metals and resources we can only imagine. This could give them the capability to build a superconductor ring or do almost anything. So perhaps in a century or two, this artificial magnetosphere will exist.

Loss of the atmosphere, if we start more of one, would take millions of years, so there is time. It lost one once before, but we weren't there to save it.

Warming Mars

It is possible to heat a planet. Most people currently believe we have already done it a little bit on Earth and I believe we have, not a lot, but a little bit. However, on Mars we really, really want to. It is a frozen tundra. Let's look at some options.

There are some ways that probably won't work. Let's start by saying the planet is in an equilibrium. Every year a lot of CO2 boils off at the South Pole. Also, every year it condenses on the poles. This describes an equilibrium whereas the summer sun heats the pole the CO2 boils off and every winter it condenses back as snow or frost. It moves away from the 6.5 mbar or so pressure and then moves back on its own. That describes a stable equilibrium, where when disturbed it naturally goes back to its previous state.

One proposal to change that equilibrium is to put a huge space mirror shining at the South Pole to increase boil off and prevent redepositing of the CO2. That probably would increase atmospheric pressure a few mbar but the other pole is just as cold. It can redeposit there. In fact, data shows that the North Pole has a lot of dust in its snow, probably from the dust storms and CO2 depositing in the southern summer when they occur. There is also the issue that current understanding of Mars is that it doesn't have that much CO2, perhaps only enough for 20 or 30 millibars, so releasing CO2 probably won't get us there. That isn't enough CO2 to affect more than a 2 to 3 degree change.

What we would need is to warm the whole planet. This is a new finding in the last 20 years or so. Previously, it was believed Mars had more CO_2.

How do we warm a planet?

Something most people don't grasp is the mass of an atmosphere. Earth's atmosphere is 5.148×10^{18} kg. That is 5 quintillion kg, or a 5 with 18 zeros behind it. Add a billion tons and you have added less than 1 part in 5,000,000,000. That is why even strong greenhouse gases we have added in large quantities like Freon are measured in parts per trillion on Earth.

Mars has a lot less atmosphere, but it still has a mass of 2.5×10^{16} kg. That is a very large mass spread over 144.9 million square kilometers. To even achieve a level of 1 part per thousand in the thin atmosphere you have to add about 25 trillion tons.

Doing calculations based on absorption relative to global warming we would have to put about 5 trillion tons of SF_6 (22,800 times the absorption of CO_2) and SF_5CF_3 (~200,000 times the absorption of CO_2, but on different wavelengths from SF_6) in order to get an increase of about 10 degrees C. There might be a small compounding effect on top of this where CO_2 and water vapor levels in the atmosphere rise as a result of the warming. However, this is not something all the factories on Earth could do if they wanted to, much less factories on Mars. These are long lived gases, as in thousands of years, but even the maintenance rate is on the order of a billion tons a year.

I read many research papers on this topic, but one in particular was looking for a way to terraform Mars. They selected a tailored group of five or six gases in order to cover a wider range of the spectrum. They then proposed releasing these gases into the atmosphere in order to achieve a seventy degree rise to match the Earth's mean temperature. I believe it would work as they claimed, but it was many times the 5 trillion tons I just suggested. So it could be done, but it might take us centuries to generate that level of greenhouse gases in Mars' atmosphere.

We could start the process if we choose to, by producing thousands or even millions of tons of these gases in factories on Mars. I am not sure the will is there to build a factory just to produce these or similar greenhouse gases, but they are used in producing solar panels and integrated circuits. Perhaps we could have 'dirty' factories on purpose in order to warm a frigid planet. That is an option. However, if we choose to do something like that it will take many years to achieve any appreciable difference.

I think for the next few decades, the answer may lay in naturally starting industries that generate some greenhouse gases. If they are released in the atmosphere, they begin the process, but very slowly. As we become more capable on Mars, we may choose to proceed more rapidly but eventually have to address the magnetosphere issue. It may take centuries, perhaps even a millennia or two, but we will probably solve it and transform Mars into an Eden. It could involve everything from space mirrors to greenhouse gases, but it won't be quick.

One final possibility is bacteria that release gases from the regolith. Generally, all bacteria on Earth require some level of liquid water, so this may be a reach. Is it possible to engineer such a bacteria? I don't know. Something biological could be a much easier answer but currently we have no life on Earth that can exist on the surface of Mars without an enclosure and an artificial environment. They can live in our greenhouses, but not on the open surface of Mars. Perhaps the answer lies somewhere in between, that we modify the environment a little and then biological forces kick in. That is beyond our current technology but may be possible someday.

For now, Mars is a little bit hard, as in very little atmosphere and radiation issues. We can deal with that, as we have described. We can colonize it. Transforming it may be a little bit slower.

Our Future

The best way to predict the future is to create it.

Peter Drucker, Management consultant

A Look Ahead

Elizabeth, a Martian colonist, writes:

"The year is 2052. We have been moving people to Mars for about 2 decades. The population of Mars is now 2,824. That includes 2,238 original colonists and 586 births on Mars, 2 of them second-generation Martians. We shut down the northern location in 2038 after we found underground water at 16 degrees latitude just east of us, near the foothills of Olympus Mons. The water was a geothermal source. It now provides us with 50 MW of power here in the first permanent Martian city, New Eden. It was named that because there is no war, no crime, and no hunger. SpaceX named it 'Love Me Tender' but we changed it. They do have a sense of humor.

"In 2038 through 2041, we set up a power grid that feeds supplemental power wherever it is needed and supports factories. This required some cabling and other resources from Earth, but we assembled it. We are producing our own plastics, metals, and concrete now and could have produced those cables if it were today. The metals, plastics, and concrete are very helpful with new structures like the underground tunnel network beneath the city and the commercial and public properties, like the New Eden shopping mall. Many of the communes had already linked themselves together with both tunnels and power lines to share and trade resources. There is an economy around this where they can share resources and barter for food or other desired commodities. Some communes are better at farming than others."

“A new geothermal location was found 42 kilometers southeast of us. That is now the start of the town of Providence. They are small but already have their own power grid, which we produced the equipment for. Eventually they may be as big as New Eden. We trade goods with them and supply them some of the things they need.

“Mars has started a new industry. We mine the asteroids. Last year, we sold $87B to Earth in high-value metals. They were mined in the asteroids, refined, and shipped to Earth, using Starships IVs. That makes everyone on the planet pretty wealthy, though we did have to give a significant cut to our parent company SpaceX based on prior contracts and their role in making all of this possible, though we have purchased 100 ships from them along the way.

“We kept about 15% of the metals for ourselves. Some is used in fuel or power production or other industries, like Ruthenium, uranium, and platinum. A lot more is going to be input for high-value finished goods we can sell to Earth. In the near vacuum of Mars, we can produce better chips and electronics than they can on Earth. This is a huge and growing market.

“We started schools a few years ago so children are not all home-schooled and parents can pursue other things. The schools include all of the normal material like reading, math, and science, but they tend to be a lot more advanced. The people that came here were mostly engineers and scientists. They were educated so their children got a head-start from their parents. Another topic covered is Mars history. That has to be a new one.

“In January (Earth time), we are expecting 832 new residents. We will welcome them and help them build their habitats and greenhouses. However, farming has become an industry. One of our residents has made hundreds of millions on the mining trade and has built over 100 acres of greenhouses. He intends to sell the food. Good for him. Industry is a boon to the economy and some people may need it. We can’t all be farmers forever. I believe he has hired about 150 people to work the farms, but they have a profit sharing plan so they benefit more than just their salary. That is definitely the norm here. Everyone works and no one is poor. There is a thriving economy.

“One day, we will be richer than Earth. That is because of the resources we have access to, here and in the asteroids. Did you know that Mars has higher percentages of ‘rare-Earth’ metals than the Earth does? Being richer than Earth may be a long time from now, but for now, we are prospering and living our lives the way we want to.”

Conquering Other Planets

Once we colonize Mars, I do not believe it will stop there. There are other interesting places. The asteroids have trillions of dollars of mineral deposits like iron and other metals. I wouldn't advise dumping it all on the market at once, but the wealth is there. We almost certainly will mine an asteroid or two if we stay on the track I believe we are on. Mining there, with no natural resources like water or oxygen might be a market for Mars, whether they are the people doing it or just providing resources to the miners. There are viable markets involved as long as transportation costs are within reason. If you can make your own fuel and there are reusable spaceships, transportation costs are reasonable.

A couple of moons of Jupiter are interesting. Ganymede, Calisto, and Europa all have water ice surfaces and are believed to have oceans under the ice. They are the size of Mercury or larger and have atmospheres. They are torn by enormous tidal forces from Jupiter. Europa especially is interesting. Its ocean is sitting on its silicate core, which would provide minerals to enrich the water. Does it have life or could it offer an environment where it might be possible for life to exist? We don't know. It is very far from the Sun, but it does have other forces warming it. Can we grow crops there? Not under natural sunlight. We would have to use LED grow lights.

Can we produce power there? Well, not solar power. It is way too far from the Sun. Any power there would definitely be nuclear. Every one of our probes that went there used nuclear power, up until a couple of recent ones with huge solar arrays producing only 4% of the power they would on Earth. Nuclear is the only option for significant power with no air and that far from the sun. The good news is it is cheap and lightweight. We can produce all the nuclear power we need without taking too much mass with us. However, the trip there would take years.

Io is also interesting. No one thinks we could live there, but it is the most volcanically active body in the solar system.

The asteroids do not have an atmosphere but are huge reserves of valuable minerals.

We may conquer other worlds. They will probably be moons. Jupiter and the rest of the gas giants may not even have a surface and if they do, it is liquid methane under enough atmospheric pressure to crush us. Their moons might be an option. There are also options like putting space stations in a stable orbit, such as Lagrange Points, but they lack all resources other than maybe solar power so they can never be truly self-supporting like Mars can. They can grow crops but there are no resources provided like soil or water. After that, we have to leave the solar system to find new worlds.

Our Future in the Stars

If we stay on a track pursuing the goal of reaching out, like what has happened recently with commercial space exploration, we will someday reach past this solar system. There are ways of going. We haven't invented warp drive yet, so the very best options are probably decades or even centuries to another star. That doesn't mean we can't. It just means it is very hard now and would involve generation ships, ships that have a sustainable environment and can support multiple generations of humans. Does that mean never?

In the middle ages, if you spoke of going to the moon, they would have thought you were crazy. It wasn't possible then. Move forward about 300 hundred years and we have done that. I hope our future continues to be as bright and that we explore Mars and then the stars.

Epilogue

We will have a colony on Mars. Why? Because we can, and at least one person with incredible power, wealth in this case as opposed to political power, wants to and has the skills or has hired the skills. That man is Elon Musk. If he didn't, perhaps someone else would. I don't know, but I am very grateful that someone with the know-how and the will and the capability wants to. Starship will work according to my math and obviously theirs. We will go to Mars and we will have colonies there. Will it be a million people by 2050 like Elon suggested in an interview? I think not, but it could possibly be in the thousands and growing by then. That is a huge start.

Some of us will live on Mars, probably not me. I am already 66, but perhaps you. Some of our children will too, including those born on Mars. Many of our grandchildren will and perhaps in a few generations the population of Mars will be a very large number. That is a very hopeful thought for the preservation of the species. If the Earth is destroyed, at least the species isn't. Or perhaps someday when the Sun is growing hotter, Mars will be a refuge for those overheating on Earth.

The most important thing is that we go. We are preserving our future for ourselves and our species. I think our future and our past also is as explorers. Our ancestors didn't come to the new world to do the same thing they had done all their lives. They came to explore new things. They were adventurers. I don't just say that for the Europeans. The original inhabitants, the Native Americans, came here across the frozen ice between Russia and Alaska millennia ago it is believed. They explored and colonized two continents. It is inherent in the human nature to explore. What is out there? Let's go find out what the universe has to offer.

References

Tunguska event:
[1] https://earthsky.org/space/what-is-the-tunguska-explosion/

Asteroid strike:
[2] https://science.howstuffworks.com/nature/natural-disasters/asteroid-hits-earth.htm

Human consumption
[3]https://ntrs.nasa.gov/api/citations/20150003005/downloads/20150003005.pdf

Dust storms:
[4] https://www.nasa.gov/feature/goddard/the-fact-and-fiction-of-martian-dust-storms

Geothermally active and liquid water:
[5] https://www.space.com/mars-liquid-water-south-pole-subglacial

Sabatier reactor use of biomass:
[6]https://uwspace.uwaterloo.ca/bitstream/handle/10012/18225/Zhuang_Yichen.pdf?sequence=6&isAllowed=y

Solar power on Mars:
[7]https://ntrs.nasa.gov/api/citations/20070010752/downloads/20070010752.pdf

Record solar performance:
[8] https://www.nrel.gov/news/press/2022/nrel-creates-highest-efficiency-1-sun-solar-cell.html

Thermal Oil
[9] https://relatherm.com/product/relatherm-uht/

Sabatier performance
[10] https://www.lpi.usra.edu/meetings/isru97/PDF/CLARK_DL.PDF

Sabatier reactor design
[11] https://www.mdpi.com/1996-1073/14/19/6175/pdf

[12] https://link.springer.com/article/10.1007/s10665-021-10134-2

[13]https://www.sciencedirect.com/science/article/abs/pii/S2212982016303614

Electrolysis 95-98% efficiency
[14] https://www.greencarcongress.com/2022/03/20220321-hysata.html#:~:text=Hysata's%20overall%20electrolyser%20system%20has,less%20for%20existing%20electrolyser%20technologies.

Why not 5 degrees
[15] https://phys.org/news/2022-08-subsurface-mars-defy-physics-seismic.html

Potential geothermal power
[16] https://nhsjs.com/2021/harvesting-geothermal-energy-on-mars-for-future-settlement/

Solar flares
[17] https://www.nasa.gov/mission_pages/stereo/news/stereo_astronauts.html

Cosmic rays exposure
http://cricket.biol.sc.edu/papers/natural/Hendry%20et%20al%202009.pdf

[19] https://en.wikipedia.org/wiki/Background_radiation#cite_note-HNBR2009-28

[20] https://marspedia.org/Cosmic_radiation

Cosmic rays and shielding
[21]https://www.esa.int/Enabling_Support/Space_Engineering_Technology/Testing_Mars_and_Moon_soil_for_sheltering_astronauts_from_radiation

[22] https://hps.org/publicinformation/ate/q10406.html

Plant growth and CO2
[23] http://omafra.gov.on.ca/english/crops/facts/00-077.htm

Consume other species in Sabatier
[24] https://pubs.acs.org/doi/10.1021/acs.iecr.1c00389

Plants not affected by cosmic radiation (at space station)
[25]https://www.ncbi.nlm.nih.gov/pmc/articles/PMC4187168/#:~:text=Space flight%20experiments%20reveal%20no%20detrimental,%2C%20called%20 gravitropism%20%5B44%5D.

[26] Sychev, V.N., Levinskikh, M.A., Gostimsky, S.A., Bingham, G.E. and Podolsky, I.G. (2007) 'Spaceflight effects on consecutive generations of peas grown onboard the Russian segment of the international space station', *Acta Astronaut*,**60**, 426-432.
Doi: 10.1016/j.actaastro.2006.09.009. [CrossRef] [Google Scholar]

Aerogel on Mars
[27] https://www.nature.com/articles/s41550-019-0813-0

[28] https://www.jpl.nasa.gov/news/want-to-colonize-mars-aerogel-could-help

Growing crops on Mars
[29] https://blogs.scientificamerican.com/observations/how-to-grow-vegetables-on-mars/

Land to feed a person
[30] https://www.growveg.com/guides/growing-enough-food-to-feed-a-family/

Crops and CO2 consumption
[31] https://climatefeedback.org/claimreview/a-2014-study-showed-that-the-us-corn-belt-is-one-of-the-biggest-primary-producers-on-earth-in-july-but-didnt-show-that-it-produces-more-oxygen-than-the-amazon/
(They were originally trying to debunk a myth but went on to provide the actual data.)

Urine as fertilizer
[32] https://www.scientificamerican.com/article/human-urine-is-an-effective-fertilizer/

Additional References

Going to Mars is a safe dose of radiation (it is a scientific paper even if some education site is posting it):
https://www.lehman.edu/academics/education/middle-high-school-education/documents/mars.pdf

Solar flux at Mars
https://ntrs.nasa.gov/api/citations/19890018252/downloads/19890018252.pdf

Aerogel properties
http://www.aerogel.org/?p=16

Aerogel in a vacuum
https://physics.stackexchange.com/questions/571216/does-aerogel-blow-up-in-a-vacuum

Solar panels only weighing 1.5 kg/m2
https://www.renogy.com/blog/lightweight-solar-panels-maximizing-solar-energy-wherever-you-go/#:~:text=Some%20of%20the%20lightest%20solar,less%20than%20200%20micrometers%20thick

Global temps for the last 1,000 years on Earth
http://www.faculty.ucr.edu/~legneref/bronze/jpg/World%20Climate-4.jpg